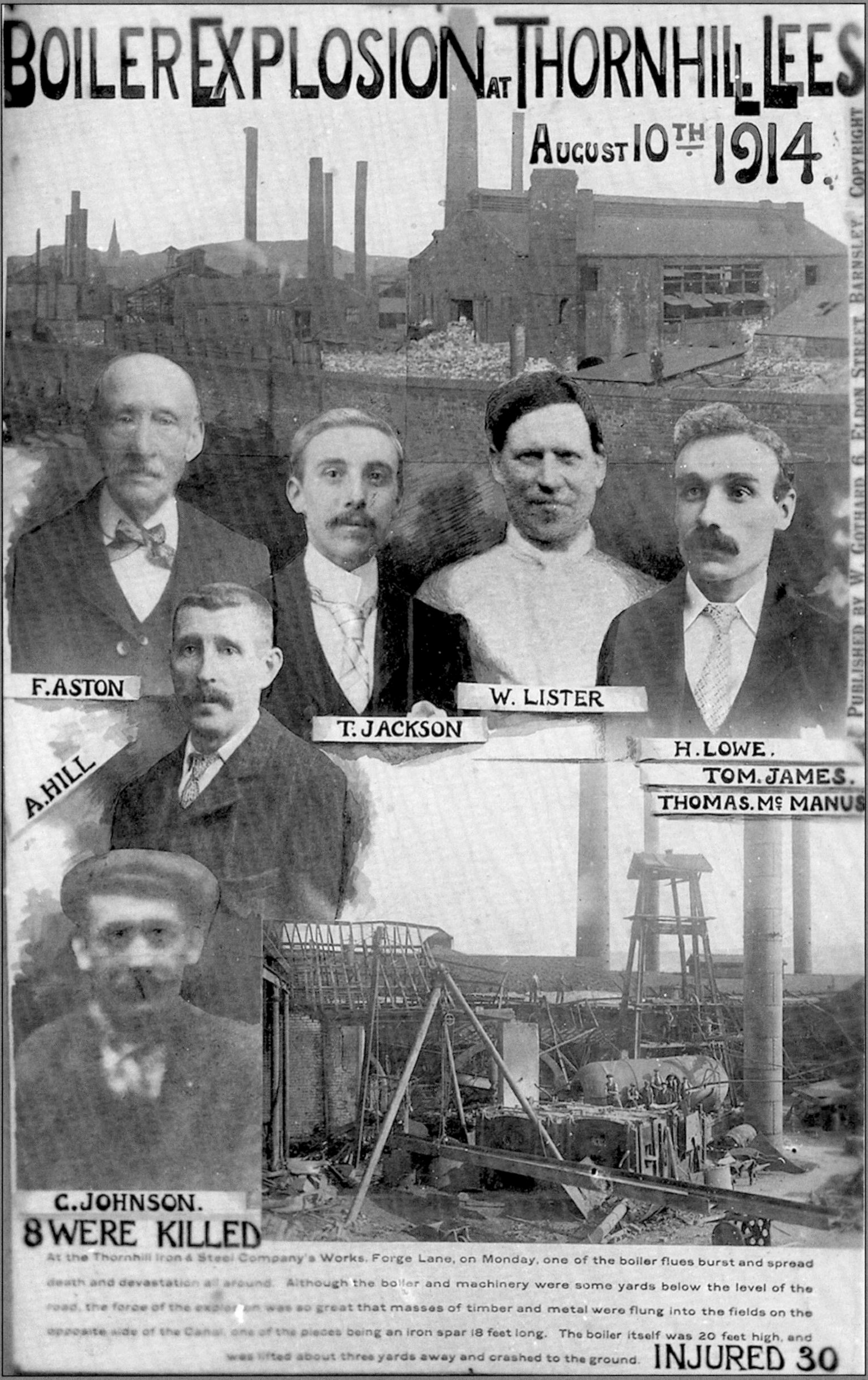
BOILER EXPLOSION AT THORNHILL LEES
AUGUST 10TH 1914.
PUBLISHED BY W. GOTHARD 6, ELDON STREET, BARNSLEY. COPYRIGHT
F. ASTON
T. JACKSON
W. LISTER
H. LOWE.
A. HILL
TOM. JAMES.
THOMAS. McMANUS
C. JOHNSON.
8 WERE KILLED
At the Thornhill Iron & Steel Company's Works, Forge Lane, on Monday, one of the boiler flues burst and spread death and devastation all around. Although the boiler and machinery were some yards below the level of the road, the force of the explosion was so great that masses of timber and metal were flung into the fields on the opposite side of the Canal, one of the pieces being an iron spar 18 feet long. The boiler itself was 20 feet high, and was lifted about three yards away and crashed to the ground.
INJURED 30

BLOWBACK

An Anecdotal Look at Pressure Equipment and Other Harmless Devices That Can Kill You!

PAUL BRENNAN

Foreword by

DR. EDWARD TENNER

International Best-Selling Author of

Why Things Bite Back: Technology and the Revenge of Unintended Consequences

Featuring First-Person Excerpts from the Rare 1892 Chester D. Berry Book

Loss of the Sultana and Reminiscences of Survivors

BLOWBACK

For information about this title or to order other books and/or electronic media, contact the publisher:
The National Board of Boiler and Pressure Vessel Inspectors
1055 Crupper Avenue, Columbus, OH 43229-1183
nationalboard.org
614.888.8320

ISBN: 978-0-9889541-0-6 (Hardcover)

Printed in the United States of America

Cover and Interior design by: 1106 Design

blowback (blō'băk') n.

1 an unforeseen and unwanted effect, result, or set of repercussions **2** the backpressure in an internal-combustion engine or a boiler **3** powder residue that is released upon automatic ejection of a spent cartridge or shell from a firearm **4** a violent release of energy *syn.* blowup, detonation, explosion

Dedicated to Those Lives Abbreviated
by
Needless Tragedy

Fifty million people die every year,
6,000 die every hour, and over 100 people
die every minute. But when thousands of
people die in the same place and at the same
time, we are more likely to wonder why God
would allow such a thing to happen.

– *Steve Farrar, American writer,* God Built, *2008*

TABLE OF CONTENTS

FOREWORD

To most of the public, "technology" means electronic devices. But how are the batteries in laptops, tablets, and smartphones charged? According to statistics published by the U.S. Census Bureau, 90.6 percent of American, and 82 percent of global electricity was generated with boilers. Matter under pressure is vital to contemporary society, but we pay too little attention to it until tragedies occur. Engineers have pointed out how many innovations have been the result of catastrophic failures. In this fascinating book, Paul Brennan illustrates how true this is of pressure containment equipment, a subject that most historians of technology, as well as lay people, have neglected.

Our very lives are the story of pressure vessels. Autoclaves sterilize the obstetrical instruments that help bring us into the world. Not only is much of our food and drink processed under pressure, but it occasionally is itself a natural pressure vessel. When growth chemicals in China were used recently to create bigger watermelons, the fruit actually began to explode, destroying 115 acres of crops. Even after death we may become risks in our own right; methane released by decomposing bodies can lead to what funeral directors call "explosive casket syndrome," so burial containers have discreet pressure relief valves.

Brennan's ***Blowback*** documents not only the ubiquity of boilers and pressure vessels in our lives, but the devastating effects of taking their safe operation for granted. It's a story of ingenuity, not least of one of the most underrated inventions of all time: the seventeenth century French physician Denis Papin's safety relief valve. Sadly, it's also a story of idiocy, as when an 1890s Texas showman staged a train crash in which the engines' boilers exploded lethally, or when today's revelers throw pressurized metal beer kegs into fires. It's a story of misplaced economy, as when the use of waste gas from nearby fields to power individual room heaters in a Texas school led to an accumulation that, once detonated, blew the school into the air, killing nearly 300 people, including many children. (The chemicals producing the now-familiar rotten egg smell of domestic natural gas were mandated after the event.) And it's the story of heartbreaking greed: The deadliest accident in American history, a boiler explosion on the Mississippi steamboat *Sultana,* resulted not just from shoddy repairs but from overloading the vessel with Union veterans of horrific Confederate prisoner-of-war camps. While other ships were available, the captain-owner, paid by the head, crammed over 2,500 people on board, of whom 1,800 perished.

Often harrowing, sometimes amusing, ***Blowback*** will give all readers a new respect for all the high-pressure equipment that underlies their way of life—not least the domestic hot-water heater. It will show how indebted we are to the American Society of Mechanical Engineers for establishing a national boiler code almost 100 years ago, and to The National Board of Boiler and Pressure Vessel Inspectors, founded in 1919, for its work in training inspectors and registering 45 million pressure items to date. It is hard to realize that at the turn of the last century, on average two people a day were still killed by boiler explosions in the United States.

Blowback reveals a symmetrical set of unintended consequences. It is often the greatest disasters, many caused by recklessness, that inspire safety programs that the best-grounded technical arguments had failed to secure. As Brennan notes, the great explosions of the 19th and early 20th centuries were directly responsible for new laws and inspection programs in Connecticut and Ohio. As recently as 2006, the state of South Carolina resisted boiler safety laws in the name of freedom from government interference until a fatal factory explosion propelled a boiler across a four-lane highway.

But there is also a potential unintended consequence of the success of legislation. Progress may lead people to take for granted the essential work of inspection, training, and certification. And new forms of pressure vessels may create fresh risks. ***Blowback*** reminds us of the constant vigilance and technical imagination that public safety demands.

—Edward Tenner, Ph.D.
International Best-Selling Author:
Why Things Bite Back: Technology and the Revenge of Unintended Consequences
Plainsboro, NJ
May 25, 2012

ACKNOWLEDGEMENTS

Dick Allison

Whose encouragement and friendship have been a genuine inspiration through years too many to calculate.

Kimberly Ball

Whose editorial pragmatism, perseverance, and attention to detail were accomplished at times of great personal challenge.

Dave Douin

For his support, vision, and leadership in reaching out to educate an unsuspecting yet dangerously vulnerable general populace.

Marilyn Hill

For her patience, kindness, sense of organization, and admirable ability to keep me pointed in one direction.

John Hoh and Jim McGimpsey

For comments gently filtered through a prism of technical enlightenment.

Judi and Ron Howard

For Judi's savvy editorial guidance and Ron's limitless technical expertise. For the delightful acquaintance of both.

Dick Neustadt

Without whose longtime association I would be less the wiser.

Brandon Sofsky

Who never fails to amaze me with his discerning visual style and eye for the unique.

The late Don Tanner

For his trust and confidence and for bringing the word "fun" back into the workday lexicon.

Joan Webster

For being . . . well, just Joan Webster.

Wendy White

Whose bountiful editorial talents continually make me look as though I know what I'm doing.

The late Warren Widenhoffer

From whom I learned there is nothing in this world that cannot be accomplished . . . if one doesn't mind who gets the credit.

My Family

My loving wife, la joie de ma vie, and best friend of 40 years, Susanne; my daughter Shauna, who possesses twice the creative talent of her old man; and my son Tyler, whose inner strength continually reminds me patience is a profound and precious virtue.

Lastly: my late parents, Paul Sr. and Virginia Brennan, without whose loving union this book—and author—would not have been possible.

PROLOGUE

ASK ANYONE WHAT WAS the most important invention leading to growth of industrialized nations, and most will answer electricity, the combustion engine, or some entity of perceived validity.

Fact is, none is remotely correct. The industrial might of great countries was born of fire and water . . . steam.

Significant to the production of steam is pressure equipment.

While there have been many books about pressure equipment, few have captured the importance of these devices in satisfying our enormous appetite for that which we consume and, unfortunately, take for granted.

If one craves human comforts such as a warm house and hot running water, this equipment is as essential as a computer. Indeed, a computer could not operate if it didn't have electric power supplied by turbines driven by mammoth steam generators.

But most people rarely give thought to pressure devices. Without these mechanisms, however, civilization as we know it today would not exist.

Pressure containment equipment such as boilers, pressure vessels, hot water tanks, and autoclaves, is innocuous when properly maintained and operated. The underlying problem affecting the general public is a lack of understanding of what can happen *if* these mechanical mainstays are ignored or inadequately monitored.

As the reader will observe, the foremost cause of pressure equipment accidents is human error. Much of what follows chronicles human missteps accompanied by tragic results.

A number of these incidents occurred in the late 1800s and early 1900s, before passage of pressure equipment laws. Unlike pressure mechanisms of that era, much of today's steam-generating equipment used in industry is regulated and professionally inspected. Other, more pedestrian containment items such as water heaters and beer kegs, although equally as dangerous, receive relatively little inspection oversight. But many of the codes and standards applicable to the construction, operation, and repair of many pressure objects can trace their genesis to organizations such as ASME (American Society of Mechanical Engineers) and The National Board of Boiler and Pressure Vessel Inspectors.

This work concerns a variety of pressure containment items, many of which are boilers. Also addressed are different assortments of pressure vessels. Some large. Others small enough to fit in one's pocket.

Size notwithstanding, most of these pressurized devices possess a destructive capability unknown to the masses. This lack of knowledge is compounded by yet another little-known axiom: Each person in the civilized world comes within close proximity of a pressure equipment item every day and sometimes several times over the course of just one hour.

While a pressure-containing item is created to advance a positive outcome, risk is directly proportional to how it is used and—unfortunately—abused.

Information and knowledge are vital ingredients in keeping an unassuming public from harm. Hence the publication of this educational, somewhat entertaining, and at times tragic narrative.

The following assortment of reflections and anecdotes covers nearly 2,000 years of an evolving engineering process, from the modest yet dangerous beginnings of pressure technology to the culmination of great industries influencing every aspect of the human condition.

Herewith a record of words and illustrations on how pressure equipment has influenced not only our daily lives, but the dreams and expectations of our ancestors.

Once he has turned the last page, it is hoped the reader will have a palpable understanding of and new respect for pressure equipment and the omnipresent potential of a tragic conclusion.

CHAPTER 1

THE TACIT TRUTH ABOUT TRAGEDY

Our lives are universally shortened by our ignorance.
– *Herbert Spencer, English philosopher, 1820–1903*

History is littered with examples of laws born of catastrophe.

And that is the little unspoken truth about tragedy: Without death and destruction, the number of safety regulations would be far fewer than the thousands upon thousands shaping our lives today.

Fact is, human longevity is directly proportional to the effectiveness of a society's commitment to safety. This thesis is exemplified by the incalculable number of tragedies in Third World countries where there exist few, if any, safety codes or standards, and where enforcement of safety regulations is under-financed, corrupted, or considered low priority compared to other human essentials.

In theory, the accidental deaths of countless human beings are not in vain: It is the collective sacrifice of life that compels improvement of existing safety disciplines. Nowhere is this better illustrated than by chronicling the evolution of the pressure equipment industry.

On Thursday, March 2, 1854, an unattended boiler catastrophically failed at the Hartford Fales and Gray Car Works in Hartford, Connecticut. It was rumored the vessel exploded after the operator abandoned his post to sneak a beer. A total of 21 people were killed with dozens injured.

While the real cause was never accurately determined, it was this particular incident that spawned the pressure equipment regulatory process.

Three years following the accident, a group of men interested in the physical sciences and steam safety founded the Polytechnic Club of Hartford. Before the club disbanded in 1861 during the Civil War, members discussed the idea of bringing together the dual functions of inspecting and insuring steam boilers into one harmonious organization.

And then the worst pressure equipment accident in history occurred—the devastating sinking of the steamship *Sultana* and the loss of an estimated 1,800 lives on April 27, 1865.

Within a year of *Sultana's* demise, two former members of the Polytechnic Club created the Hartford Steam Boiler Inspection and Insurance Co. (HSB). Focus of the new organization "was not insurance as such, but safety itself."

While HSB established higher standards for boiler construction and inspection, the number of pressure equipment-related deaths continued unabated.

Grover Shoe Factory: After the explosion.

On March 20, 1905, a total of 58 lives were taken when an unattended boiler overheated and exploded at the Grover Shoe Factory in Brockton, Massachusetts. The horrendous blast launched the boiler through the roof of the four-story factory with such force that it leveled the entire building.

As a consequence of this avoidable accident, the commonwealth of Massachusetts enacted the most rigorous boiler inspection laws in the United States. Today, more than 100 years later, these laws remain the toughest in North America.

At a 100-year commemoration of the Brockton disaster, a letter was read from Donald Tanner, Executive Director of The National Board of Boiler and Pressure Vessel Inspectors. He wrote:

> *This morning, you will recount many of the terrible details of this horrible human tragedy. And what you will undoubtedly learn is that the precious lives of these victims were not lost in vain. It is entirely appropriate to surmise their sacrifice prevented subsequent death and serious injury to hundreds, if not thousands, of future Massachusetts citizens—some of whom might be among you this morning. The commonwealth of Massachusetts not only has an effective boiler law, it possesses one of the best safety records in North America. The tremendous release of energy from a boiler explosion is beyond the comprehension of many. As demonstrated on March 20, 1905, some boilers possess the capability to level an entire city block. But today, 100 years later, I can state without reservation that each of you can sleep well this evening knowing the state is doing everything within its considerable power to preserve the safety and well-being of all Massachusetts citizens. Hopefully, every one of us has learned that when it comes to safety, there are no second chances.*

While Massachusetts may have raised the bar of safety, pressure equipment accidents still occurred at an unrelenting pace. Nothing, not even the cost of human lives, would get in the way of the industrial revolution.

Five years after Brockton, three boilers simultaneously failed at the American Sheet and Tin Plate Co. in Canton, Ohio. The explosion on May 17, 1910, resulted from an operator adding cold water to one of the overheated boilers.

One hundred workers were in the plant at the time of the blast. Fragmented pieces from the building, as well as body parts, were launched as far as 600 feet from the explosion site. The body of one worker was blown completely through a house before coming to rest on a neighbor's fence.

The Canton catastrophe killed 17 individuals. An additional 50 were injured.

At the time, Ohio had no boiler laws. More important, Ohio was reported to be a

Postcard photo of house through which the body of a worker from American Sheet and Tin Plate Company entered (at second story square at side of house) and exited before landing on the fence of a home across the street (behind man with bicycle).

dumping site for old boilers coming in from surrounding states.

The following year, a boiler inspection act was passed by the Ohio General Assembly and signed into law. In 1913, the state formed the Industrial Commission of Ohio, an agency that exists to this day which is responsible for the inspection of pressure equipment.

Ongoing deaths from a somewhat untested boiler technology of the 1800s left the American engineering community troubled. But an organization, founded in 1880, would prove instrumental in establishing technical standards forever impacting boiler safety.

That association, the American Society of Mechanical Engineers (ASME), adopted its boiler code in 1915. According to ASME:

> For want of reliably tested materials, secure fittings and proper valves, boilers of every description, on land and at sea, were exploding with terrifying

> frequency.… Engineers could take pride in the growing superiority of American technology, but they could not ignore the price of 50,000 dead and 2 million injured by accidents annually.

While the ASME boiler code provided a solid reference for construction standards, it lacked an important component: the authority to regulate. This need was complicated by the existence of local and state jurisdictions each having their own codes and standards. The result was a patchwork of confusion and inconsistency.

On December 2, 1919, Ohio chief inspector C.O. Myers met with chief inspectors from other jurisdictions to discuss creating a board comprising inspector representatives from each of the existing jurisdictions. That board today is known as The National Board of Boiler and Pressure Vessel Inspectors and it is the bond through which codes and standards are integrated among differing jurisdictions.

There are no longer thousands of deaths caused by pressure equipment failure each year. However, if not properly maintained and inspected, boilers and pressure vessels can still be lethal, and, in some instances, catastrophic.

The first meeting of Ohio boiler inspection officials took place August 16 and 17, 1916, in Columbus, Ohio. C.O. Myers, who would be instrumental in developing a national organization of jurisdiction inspectors, is pictured in the front row fourth from the left.

For example, rupture of a typical 30-gallon home hot-water tank generates the equivalent of 0.16 pounds of nitroglycerin. Translated, that is enough force to send the average car (weighing 2,500 pounds) to a height of nearly 125 feet—higher than the elevation of a 14-story building, with a starting lift-off velocity of 85 mph.

When such a hot-water tank explodes, its volume expands by a factor of approximately 1,600. That is comparable to a 5-gallon trash can filling a 12' x 11' x 8' living room in a split second.

While considerable progress in boiler safety has evolved since the early 1900s, explosions have continued to occur periodically.

On October 3, 1962, a boiler operator at the New York Telephone Co. in downtown New York started his boiler before leaving the building for lunch. The unattended boiler notwithstanding, there were also a number of mechanical issues contributing to an explosion killing 23 and injuring 94. Among these: non-operational safety systems, a missing mercury switch, and a pair of nonfunctional safety valves.

Concussion from the blast launched the boiler forward about 120 feet, through concrete walls and toward a nearby cafeteria, where dining employees were killed and disfigured.

One of New York's deadliest accidents resulted in enactment of a new state low-pressure boiler law.

Many victims of pressure equipment accidents have been adults, most of whom worked in or around places of business. But there have been countless child deaths. Unfortunately, a number of these accidents occurred at schools in session.

On October 13, 1980, at the Gate City Day Care Center in Atlanta, preschool-aged children

1ST SCHIRRA PHOTOS

See Page 5

WEATHER
Tonight: Cloudy, drizzle, in 60's.
Tomorrow: Cloudy, some rain, 70.
SUNRISE: 6:55 A.M.
SUNSET: 6:34 P.M.
SUNRISE TOMORROW: 6:56 A.M.

New York Post

Re-entered as 2nd class matter Nov. 22, 1949, at the Post Office at New York under Act of March 3, 1879.

Vol. 161 No. 271 NEW YORK, THURSDAY, OCTOBER 4, 1962 10 Cents

LATEST STOCK PRICES
Pages 59-62

Where 21 Died ↓

WHY?

Fire officials inspect boiler wreckage. See Pages 2 and 3. Post Photo by Pomerantz

Four Probes Into Phone Blast

Death scene: Gate City Day Care Center, Atlanta, Georgia.

played without a care in an environment their parents thought to be secure.

That all took a deadly turn as the boiler was started for the first time in anticipation of approaching cold weather. In a blink of an eye, the cast-iron boiler exploded with a force that left four children and one adult dead, and seven children seriously injured.

Reports would later explain the boiler was only partly filled with water. However, the low-water cutoff mechanism had not been wired into the system, thus rendering the protection device useless.

Public fallout from the Gate City Day Care Center tragedy provoked Georgia lawmakers to begin efforts to make certain such a heinous event would never recur. In 1984, the Georgia Legislature passed the state's first Boiler and Pressure Vessel Act.

Barely 14 months later, disaster struck an elementary school in Spencer, Oklahoma, when an 80-gallon water heater exploded in the school's cafeteria kitchen. January 19, 1982, was the day six children and one adult were killed at the Star Elementary School. Forty-two others were injured.

The water heater had been in need of repair for three years. An examination revealed evidence of controls tampering, removal of a temperature probe, and improper installation of a pressure relief valve as factors causing the accident.

As in the case of New York, Oklahoma did not have a low-pressure boiler law at the time of the horrible occurrence. Consequently, the water heater was exempt from inspection.

Nine months later, the Oklahoma Legislature passed more inclusive safety laws, covering all water heaters and boilers. The legislation provided for annual inspections as well.

Usually, accidents provoking new safety legislation involve multiple fatalities.

But in the spring of 2005, it took only one heartbreaking death to convince South Carolina legislators that resisting passage of a boiler law was becoming a liability no one could justify.

Settled in 1670 and having entered the Union in May 1788, South Carolina has long been adverse to excess legislation. And so for years, the state resisted all attempts to adopt any laws perceived to adversely affect farming and the textile industries. And that included boiler and pressure vessel laws.

During the 1975–1976 legislative session, Senator A.S. Goodstein introduced the state's

Explosion site: lunch boxes were among the few items left standing in the Star Elementary School cafeteria. The medical examiner explained all of the victims died instantly from broken necks when the blast hurled them across the room.

first pressure equipment law. It would be one of 20 introduced by 27 sponsors over the next three decades. Over 14 legislative sessions (save 1983–1984), each bill was submitted and summarily dismissed. It was not until the 2005–2006 session that a proposed pressure equipment law made it out of legislative committee.

Passed by the Senate and subsequently submitted to the House for consideration, there subsequently occurred an incident that would forever alter the Palmetto State's attitudes on boilers.

At around 10:30 P.M. on March 30, 2005, 47-year-old boiler operator Tommy Jarvis called his wife from his job at a duct and masking tape manufacturing plant in Columbia. Although most of his evening calls were pleasant discussions about the day's activities, Jarvis on this evening expressed concern about a low-pressure alarm originating from one of two boilers he maintained.

Finishing the call about 20 minutes later, Jarvis returned to his task of shoveling ash from a coal-fired boiler.

Fire investigators theorized that as Jarvis prepared to enter the firebox, an accumulation of gas in the firebox ignited when oxygen rushed in from the opened door. The ensuing explosion launched the dump-truck-sized boiler and burning shrapnel out of the plant and across a nearby four-lane highway. Jarvis died at the scene.

The spectacular accident became statewide news overnight. Legislators who heretofore questioned the necessity of pressure equipment safety laws were suddenly quieted. At a committee meeting to consider the Senate-passed legislation, representatives were seriously moved by testimony presented by the Jarvis family.

On the evening of May 17, 2005, just 75 days after introduction, S. 581, the South Carolina Boiler Safety Act, was delivered to the governor. His options: Sign the bill into law, veto the bill, or let it become law without his signature. Although Governor Mark Sanford was adamantly opposed to any new regulations, he maintained his ground by refusing to sign the bill, thus ensuring implementation of South Carolina's first pressure equipment safety law on January 1, 2006.

Thirty years following introduction of the first proposed law, South Carolina was now in harmony with other North American jurisdictions. Perhaps more important, it was in harmony with the motto adorning the great seal of South Carolina: *Dum spiro spero.*

"While I breathe, I hope." ❁

CHAPTER 2

FIRE AND WATER AND THE INDUSTRIAL REVOLUTION

The first thing to do in life is to do with purpose what one purposes to do.
– Pablo Casals, Spanish musician, 1876–1973

If the Industrial Revolution of the late eighteenth and early nineteenth centuries was a tonic for far-reaching growth and financial security, surely steam was the fountain that quenched international thirst for technology. Indeed, without steam it is reasonable to deduce this age of unparalleled growth and prosperity might have sputtered rather than spawned a new generation of economic opportunity.

There can be no doubt the era was a byproduct of an invention developed nearly two millennia earlier: the steam engine.

Creation of the first steam mechanism during the first century A.D. was credited to Heron of Alexandria. A mathematics and engineering intellect,

Heron (also referred to as Hero) invented numerous devices that remain part of present-day culture. Among these was the first vending machine (to dispense holy water), an odometer, a compressed-air fountain, a machine to cut threads into wooden screws, and what might be considered the very first automatic garage door opener (actually an opener for temple doors back then). The Greek inventor living in Egypt was also recognized as having been among the first to harness wind power through his construction of the windwheel, a device that operated a wind-driven organ.

Hero's genius was truly without equal. But it was his aeolipile that cultivated a new understanding of water heated by fire.

The word *aeolipile* is derived from the Greek *aeolos* and *pila*. Translated, it means the "ball of Aeolus." Aeolus is the Greek god of wind.

Simple in construction, the aeolipile was an enclosed round chamber containing water and suspended over fire. Conversion of water to steam forced the vapor through two crooked projecting nozzles. Emission of steam from the bent outlets caused the ball to rotate.

Interestingly, the wind ball was not unlike the operational concept of today's jet engines. The thrust (i.e., rocket principle) causes torque, which spins the ball (Newton's Third Law).

The remarkable aspect of Hero's creation: It was developed as a toy and not a mechanism of practical application. Ironically, the aeolipile was invented at a period in time where its utility, unfortunately, would be limited. And without appreciation.

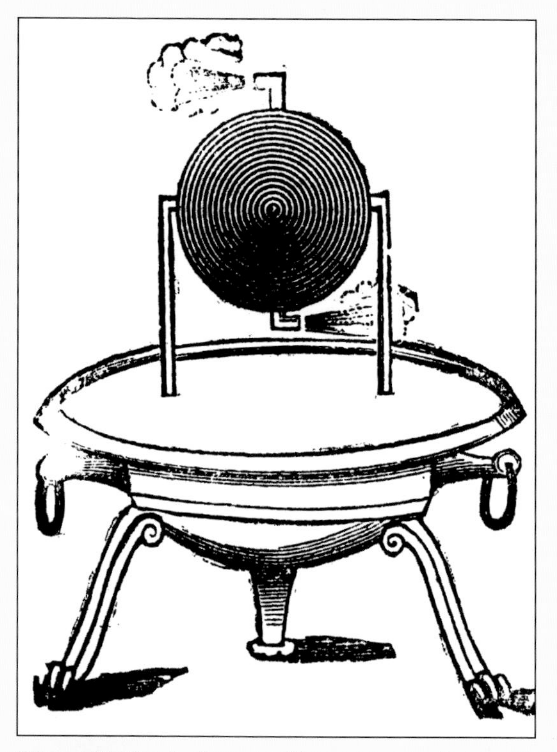

The aeolipile, the first working steam engine, was developed for inventor Hero's amusement. This illustration details how escaping jets of steam propelled the ball.

Why?

The Greeks clearly had no use for mechanical performance of work. An abundance of slaves was always available to undertake whatever task necessary. And so the potential for Hero's "toy" went unrealized. Without incentive to develop labor-saving devices and having a surplus of labor, serious application of the Greek inventor's wind ball principles would not occur until the early 1600s.

It was then need for fossil fuel achieved a new plateau.

British glassblowers required coal—lots of coal—to feed hot furnaces. But removing water from the mines was a time-consuming and terribly inefficient process. Realizing there must be a better way than using horses to carry water out of the pits, scientists began experimenting to find a better procedure.

Late in the seventeenth century, a British inventor patented a pump powered by steam. Patent holder Thomas Savery called his machine an "engine to raise water by fire."

Although the concept of using steam to create a vacuum and thus pull water upwards through a pipe had been around for centuries, it was never successfully employed. One reason may have involved safety. The cycle necessitated steam pressure provided by a boiler. Structural integrity of boilers in this period was considered inadequate, thus making the vessels subject to frequent explosions.

About 15 years after Savery recorded his patent, blacksmith Thomas Newcomen fine-tuned the boiler design with pistons and cylinders. The atmospheric engine was the first steam engine to demonstrate a positive functional value. Even though it was inefficient by today's standards, Newcomen's invention

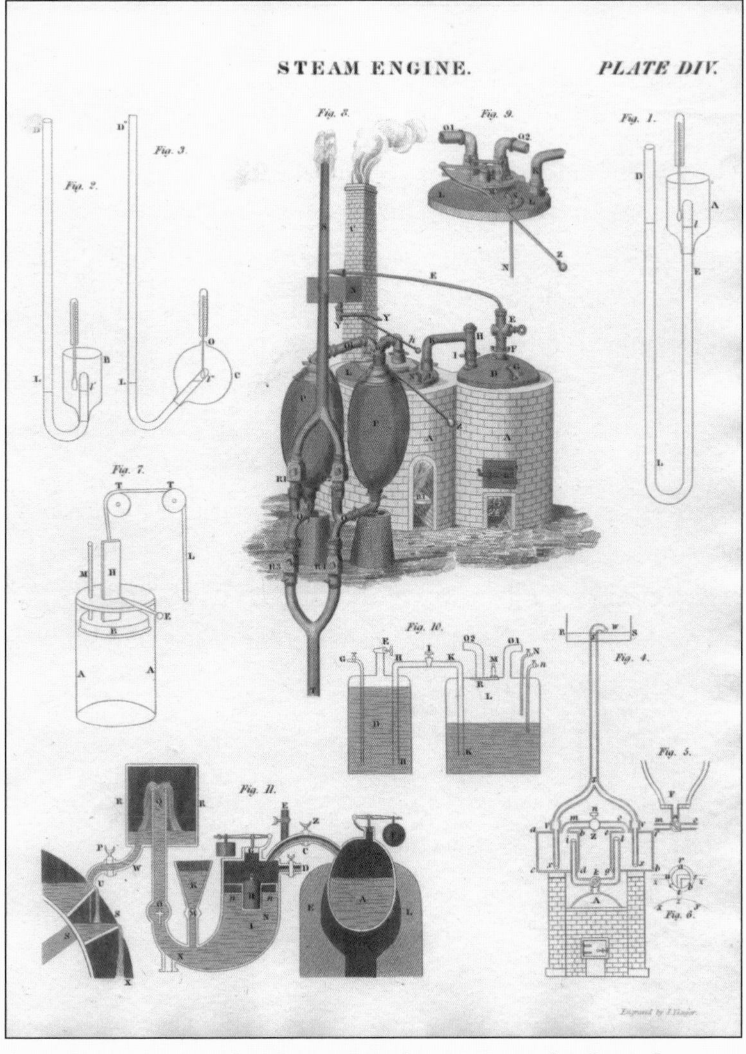

Savery's engine introduced pressurized steam into a closed vessel filled with water. As the water rose and was consequently expelled, a sprinkler condensed the steam, thus producing a vacuum capable of extracting mine water through a valve below. Savery employed two containing vessels to maximize the effect. Engraving by C. Tiebout.

did extract water from mines at depths heretofore unreachable. Construction technology that impeded development of Savery's engine apparently became a non-issue with Newcomen's design, primarily because the latter's steam was only slightly above atmospheric (low) pressure.

Perhaps the most famous name associated with development of steam technology—James Watt—entered the picture during the 1760s. As a result of Watt's marketing acumen and the efficiency enhancements he incorporated into Newcomen's invention (condensing steam in a separate compartment rather than the cylinder), the steam engine's capabilities became well-positioned for the Industrial Revolution.

Commencing in 1775, Watt's collaboration and partnership with British manufacturer Matthew Boulton resulted in production of steam engines for a variety of industries. And there were many. So many in fact, Messrs. Watt and Boulton accrued tremendous wealth that reinforced their reputations and research credentials.

But to market his products, Watt needed a dramatic yet believable selling point. So he shrewdly calculated the number of horses each engine would replace, hence minting the term "horsepower."

As for steam's role in the Industrial Revolution, historians point to the limitations on manufacturers who had to locate their facilities close to sources of wind and water power.

Oil painting of James Watt by Sir Thomas Lawrence.

Steam engines, they emphasize, permitted factories to locate anywhere, not just near rivers or large bodies of water. Consequently, factories ventured from heavily populated urban areas to the more rural. Possibilities were limitless. The revolution was on.

As the eighteenth century progressed, an interest in "strong steam," or steam at higher pressures, entered the public debate as a way to improve transportation. While an advocate of steam power, Watt expressed reservations about "strong steam," citing suspect metallurgical composition and unsafe boiler technology.

Photo of James Watt's workshop in Heathfield, Birmingham, circa. 1895.

which experienced severe boiler explosions (thus vindicating Watt's safety reservations).

As time evolved, Watt was able to decrease the size of his steam engines. Significance of a more-compact engine was not lost on other inventors. Envisioned were vehicles that could be propelled by the small, yet powerful, steam-driven mechanisms. What may not have been recognized at the time was the engine's profound effect on cargo transport, personal travel, military strategy, and virtually all aspects of modern civilization.

During the 1800s, steam use for transport figuratively exploded. By 1804, Britain ran the first steam-powered locomotive. In 1807, Robert Fulton powered the first commercial steamboat with a Watt steam engine. Shipping cargo and people via water was no longer a difficult proposition.

Never before had goods and people been mechanically moved over land and sea. Manpower and animal power were no longer the only options available.

Steam-driven transportation was not new to the 1700s. In 1769, Nicolas-Joseph Cugnot publicly displayed his "fardier" (steam wagon). Considered somewhat a novelty, the steam wagon could only travel a few hundred meters before requiring more water for its small boiler. Some historians designate Cugnot as inventor of the very first automobile.

The time period saw a number of new transportation proposals in Britain. Unfortunately, many would never see construction due to the powerful efforts of Watt and Boulton.

Strong steam advocate and cylindrical boiler pioneer Oliver Evans designed a number of steam-driven boats, some of

RELIEF IS ON ITS WAY

It is impossible to adequately discuss the advent of pressure equipment without a nod and tip of the *chapeau* to Denis Papin.

Not only did the French physicist invent what is arguably the most important component on a pressure vessel, the safety relief valve, he played an integral role in the development of the first steam engines. Some say his contributions to pressure technology were the most important since Hero.

Monsieur Papin was born in Blois, France, in 1647. Having graduated from the University of Angers, it wasn't long before the newly minted medical doctor revealed a predisposition for mechanical science.

A move to Paris allowed him to begin work as an assistant to prominent Dutch mathematician and physicist Christiaan Huygens.

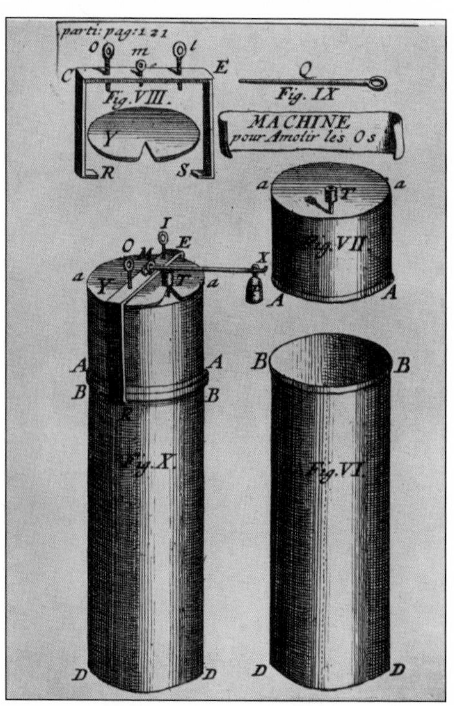

Early steam digester illustration.

Under Huygens' tutelage, Papin constructed an air pump with which he would perform numerous experiments. During this time, Papin became interested in creating a vacuum to generate motive power.

The French doctor had considerable mechanical proficiency and in 1675 traveled to London to work with one of history's most influential Anglo-Irish scientists: Robert Boyle. A chemist who refused to assume a conventional role in controlled experimentation, the English aristocrat is best known today for Boyle's Law (i.e., volume of gas varies inversely with pressure).

Papin's three years under Boyle saw him invent the steam digester in 1679. Precursor to the modern pressure cooker and autoclaves used in hospitals, the high pressure digester extracted fat from animal bones to make them brittle and consequently easily ground into bone meal. Some believed Papin's invention could soften if not liquefy stones.

The digester was essentially a closed vessel with a securely fitted lid. Contents were brought to high temperatures by increasing pressure. It is believed several of the early digesters exploded before Papin created a safety valve and employed it for the first time. This critically important relief mechanism was designed simply to relieve excessive pressure created within the vessel.

In a classic example of invention begetting invention, Papin's observance of the safety valve's up-and-down motion inspired his idea of a piston-and-cylinder engine. Perhaps for want of funding, the Frenchman never brought his design to fruition. Yet in 1697, Papin's designs were used by Thomas Savery, builder of the world's first steam engine.

The doctor turned mechanical engineer worked on a variety of other inventions during his career. Perhaps the most notable: the first steam propelled paddle boat. It was September 24, 1707, when Papin and his family boarded the ingeniously assembled boat en route to the North Sea by way of Germany's Fulda and Weser rivers. In Hannoversch Münden, he was set upon by boatmen objecting to the paddle craft's encroachment into their exclusive navigational region. His boat confiscated, Papin later made his way to London, where he arrived penniless.

Despite having developed designs for a centrifugal pump, several machines for lifting water, steam wagons, and a furnace to melt glass, most of Papin's ideas were never realized during his lifetime. He died in obscurity around 1712.

While Papin is today credited with having worked on numerous steam-related inventions, it is the safety relief valve that today endures as the French inventor's most significant contribution to mechanical technology.

The first steam tramway locomotive, invented by Richard Trevithick, was actually designed to operate on land (despite the use of rail tracks developed by the Greeks almost 700 years before Hero's aeolipile). In 1814, George Stephenson's "Blücher" was tested on a 450-foot track pulling eight filled coal wagons. Historians cite the "Blücher" as being the first steam engine locomotive to operate on a railroad. But it was not until 1825 the first public locomotive route went into operation: the Stockton and Darlington railway built by Stephenson.

Steam prompted a number of inventors to ponder its transportation uses.

In 1867, American Sylvester Roper developed a steam-driven motorcycle (the gas-powered engine was yet to be invented). Considered the world's first motorcycle, this pedal-less two-cylinder vehicle was powered by coal. An 1894 version of Roper's motorcycle resulted in the inventor's death in 1896. While riding his vehicle as a pacesetter at a bicycle race, Roper became excited by the motorcycle's "remarkable speed." He subsequently suffered a heart attack, thus becoming the world's very first motorcycle fatality.

In 1873, steam power replaced horse power to drive cable cars in San Francisco. While a long cable under city streets was used to pull the cars about the city, it was a massive power station steam engine equipped with large pulleys and wheels that kept the cable (and riders) moving. In the late 1880s, some cities turned to streetcars to transport riders. Although powered by electricity, enormous steam engines were still required at streetcar power stations to turn large generators.

While steam seemed ideal for ground transportation, there were also attempts to power flight. One individual fascinated by the possibility of flying was Alexander Graham Bell.

Rare 1825 photo of the No. 1 Stockton and Darlington Railway locomotive engine.

According to a recently published biography, Bell marveled at the ease with which birds could soar over the countryside. Also realizing the dangers of flying, he decided in the late

1800s to construct a small device in his laboratory bearing semblance to a helicopter.

Designing a tiny boiler weighing less than a pound, Bell had his assistant William Ellis construct the steam mechanism from tin. The boiler was connected to a vertical pipe on which was placed a double blade propeller made of cloth.

Bell's plan was simple: "an upright tubular boiler could be made to lift itself in the air—fuel and all—by fan wheel arrangement worked by a simple jet of steam," he later wrote.

Building pressure in the miniature boiler was not easily accomplished but achievable. And when it did occur, the mechanism shot across the lab, propelled by steam escaping from two small holes. Failing to achieve the desired result, Bell abandoned his efforts but never discarded the notion man would someday take flight.

The rest, as they say, is history.

We know the first Industrial Revolution started in the late eighteenth and early nineteenth centuries. And we know of its profound impact on agriculture, manufacturing, and transportation. But we also know steam power played a significant role in the evolution of what we now know as the civilized world.

Some historians claim the Industrial Revolution was inevitable and gestating long before steam equipment took front and center on the world stage. While the impetus for industrial development may have been provoked by world socio-economic events, it can be validated steam was *a* driving force if not *the* driving force in accelerating the revolution.

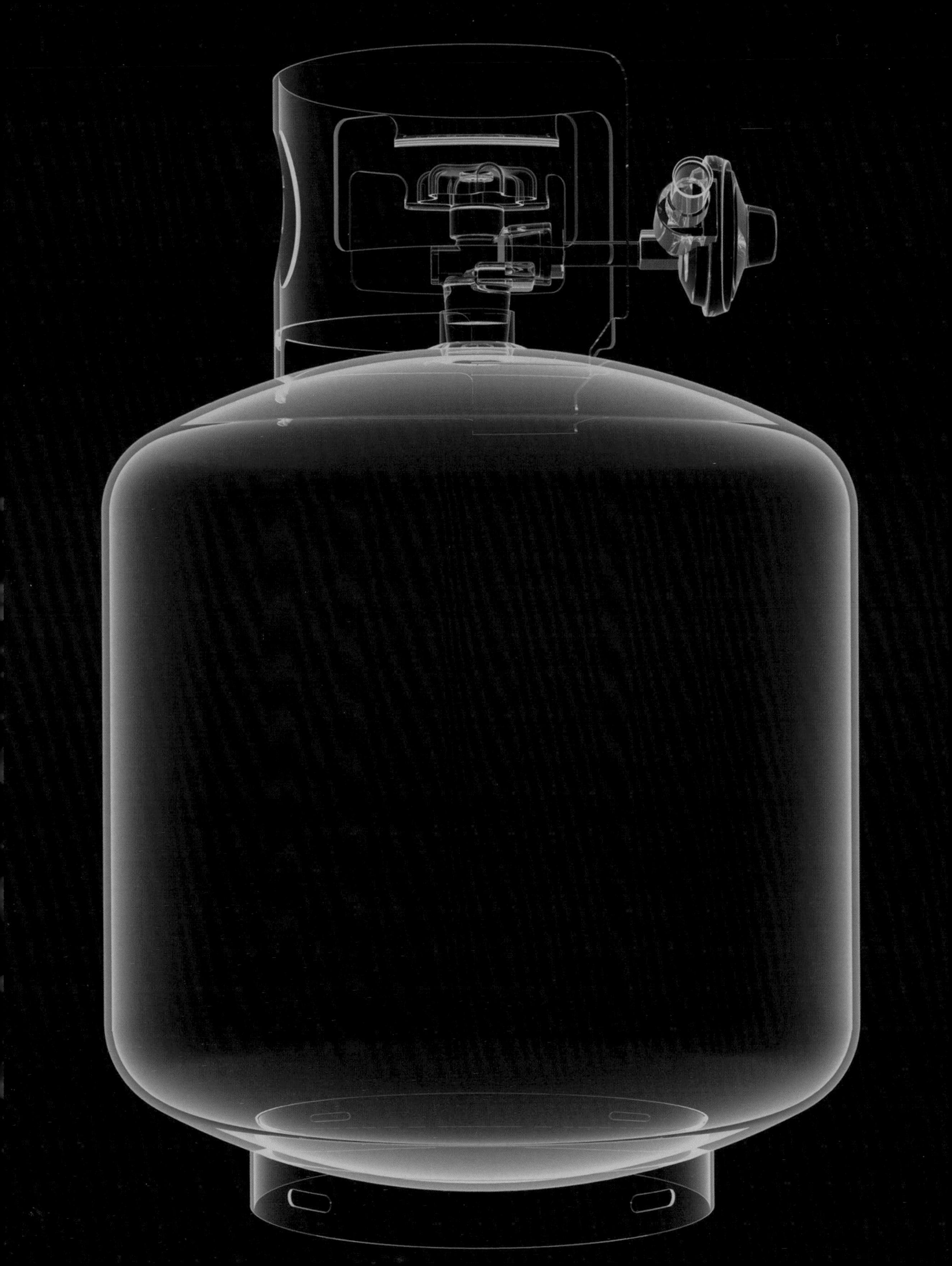

CHAPTER 3

SO . . . WHAT IS A PRESSURE VESSEL?

Pressure is something you
feel only when you don't
know what you are doing.
– *Chuck Noll, football coach*

For the purpose of simplicity, let's just say a pressure vessel is a boiler or storage tank capable of withstanding pressures in excess of 15 psig exerted by enclosed matter.

This pressure comes in a variety of forms: steam, gaseous matter, air and liquid.

Pressure vessels are essential in the manufacturing process industry because they contain various gases and liquids—such as chemicals—stored under high pressure. (Smaller containers also have widespread applications in the private sector as well as personal use as in dispensing medicines, cooking, cleaning, etc.)

Industrially, pressure vessels are today designed and manufactured to certain strength parameters for the purpose of precluding rupture and an ensuing explosion.

In the early 1900s, standards addressing design, manufacture, and operation were developed by ASME and The National Board of Boiler and

Pressure Vessel Inspectors. These standards were subsequently adopted as laws administered and regulated by North American jurisdictions to address vessel strength and maximum safe operating pressures. Major attention focused on manufacturing to ensure each container was designed for specific conditions and usage.

Although the lay person is seldom exposed to giant industry generators and massive chemical storage containers, one should understand pressure vessels are found everywhere: from hot water heaters in nearly every building in the civilized world to the seemingly innocent pocket inhaler used by asthmatics.

Typical examples of pressure equipment include:

- Tanks containing liquefied gases, such as ammonia, chlorine, propane, butane, and LPG
- Compressed air receivers
- Domestic hot water heaters
- Sterilizers
- Autoclaves
- Diving tanks
- Distillation towers
- Airbrake reservoirs on road vehicles
- Submarines
- Space vehicles
- Space suits
- Diving suits
- Recompression chambers
- Cryogenic tanks
- Chemical tanks
- Pressure cookers
- Digesters
- Petrochemical tanks
- Hydraulic pumps on airplanes
- Steam catapults on aircraft carries
- Tennis balls
- Aerosol dispensers
- Tanker trucks
- Storage tanks
- Heat exchangers
- Machine tools
- Aluminum baseball bats
- Fire extinguishers
- Food and beverage tanks

This pressure equipment generally can be found in:

- Mines
- Refineries
- Chemical facilities
- Nuclear reactors
- Schools
- Restaurants
- Churches
- Hospitals
- Rest homes
- Dry cleaners
- Car washes

- Automobiles
- Stadiums
- Assembly halls
- Petroleum refineries
- Petrochemical plants
- Pharmaceutical plants
- Food processing plants
- Cement plants
- Pulp and paper plants
- Food plants
- Beverage plants
- Fossil fuel and nuclear power plants

While both of these lists are far from exhaustive, it is reasonable to assume that while you can perhaps *hide* pressure vessels (e.g., behind walls), you cannot hide *from* pressure vessels. ⚙

CHAPTER 4

THE ROOTS OF PRESSURE VESSEL DEVELOPMENT

Because of this, originality
consists in returning
to the origin.
– Antonio Gaudi, Spanish
architect, 1852–1926

Although man developed and refined the pressure vessel, it was nature that served as inspiration. In addition to providing excellent and elementary examples, it is nature that best illustrates the concept of containing pressure.

Example: a popcorn kernel.

A seed, the kernel comprises a hard shell containing a miniature plant embryo surrounded by starchy material providing nutrients and water for the growth process. Exposed to high heat (400°F), the water is converted to steam, which causes the kernel to explode and yield its tasty cargo. The eruption actually turns the kernel inside out to a size approximately 40 times its original mass.

If not for the amazing resiliency of the—metaphorically speaking—outside pressure vessel (i.e., hard shell), which protects and tightly seals its contents, our enjoyment of corn would be limited to only that on the cob.

How resilient?

The oldest popcorn ever discovered was about 5,600 years old and found in the "Bat Cave" of central New Mexico. So well preserved were the ancient kernels that they could still pop.

More recently, researchers found what they described as surprisingly fresh popcorn kernels estimated to be at least 1,000 years old.

Yet another example of nature's pressure vessel is the watermelon. *Citrullus lanatus*.

This delectable summer fruit sort of grows its own pressure vessel. And while this green integument keeps the internal produce juicy and tasty, it differs from a manufactured containment vessel in that it lacks a pressure relief device or safety valve.

So why would a gourd need a safety valve? How much pressure can sweet, red, watery pulp exert?

Ask farmers near Danyang in eastern China, who in the spring of 2011 witnessed the exploding watermelon phenomenon first hand. The cause, according to Environmental Protection Agency sources, was forchlorfenuron, a plant growth regulator farmers sprayed on their crops to accelerate growth.

Neighbors humorously referred to the fruit eruptions as "land mines." But the 20 or so affected farmers were not amused. Approximately 115 acres of the gourds were destroyed and eventually fed to pigs and fish.

While normal watermelons don't explode, introduction of the growth chemical stimulated the fruit to expand rapidly and thus create internal pressure. When that pressure exceeded the rind's containment capacity, the melon—as any pressure vessel—ruptured.

Unlike some other pressure vessel breaches, however, it was fortunate no one was injured.

At least no one admitted to being injured by an exploding watermelon.

Melon mayhem: Chinese farmer surveys damage.

NOWHERE TO RUN

Why does it take a minute
to say hello and forever
to say goodbye?
– *Unknown*

It has been reported a large industrial pressure vessel has the energy capacity to level an entire city block.

But in 1944, an explosion of a liquefied natural gas tank actually brought about the destruction of more than one square mile on Cleveland's east side.

During the 1940s, above-ground storage of natural gas was a common practice in buildings, homes, and manufacturing facilities across America. The East Ohio Gas Company tank farm located on East 61st Street close to Lake Erie was such a facility.

On Friday, October 20, at about 2:30 p.m., a seam breached on the side of tank number four. (It is believed an alloy used to construct the tank could not contain the liquid gas' cold temperature.) As a result, the 900-million-cubic-feet-capacity storage vessel began to emit liquefied natural gas vapor. Lake Erie winds floated the vapor toward the east side Cleveland community and into sewer lines by way of street catch basins.

The combination of liquefied natural gas, air, and sewer gas subsequently ignited (source of the ignition could not be identified), thus launching an inferno that instantly erupted through the area's sewer system. As flames shot out into the streets, manhole covers took off like flying saucers in a low-budget science fiction movie. Investigators would later locate one of the covers several miles east of the scene in the Cleveland community of Glenville.

The incident prompted an exodus from homes and businesses in the area. Shortly before three o'clock, however, residents began a return to their homes with the impression firemen had the situation under control.

But at three o'clock, yet another tank exploded. It was a blast that would level the entire tank farm.

Those who ventured back to their homes were suddenly trapped as fire and explosions precluded their escape. Houses covering a 20-block area became engulfed in flames as fire traveled the length of sewer lines and up through drains in sinks. Those still at home reported clothing catching fire right after the second explosion.

Initial death estimates came in at around 200. But Cuyahoga County's coroner added a caveat: The intensity and volatile nature of the explosions and consequent fires might have been enough to vaporize human flesh and bone. It would be weeks before an exact death count could be tallied.

Although the death toll was later reported to be 131, the disaster left in its wake a much wider swath of destruction. A total of 225 were injured, and 600 people were rendered homeless. Seventy-nine homes, two factories, 217 cars, and seven trailers were also destroyed, as well as much of the underground infrastructure and utility system. *Time* magazine referred to the incident as a "major catastrophe of the modern industrial era."

Experts said the number of deaths would have been much higher had children not been at school or adults at work.

Many of those affected reported tremendous losses of personal assets including cash, stocks, and bonds (many people kept such valuables at home following the Great Depression). Some lost everything they owned. Estimated personal and industrial property damage totaled between $7 million and $15 million ($91 million and $196 million respectively in 2012 dollars).

East Ohio Gas Company assumed responsibility for the tragic turn of events. It provided direct financial assistance to the needy and rebuilt the community.

While the Cleveland East Ohio explosion exacted an incredible toll, there was one positive outcome: Utilities as well as communities began to re-evaluate the safety of aboveground natural gas storage systems. Not long after, tankless below-ground storage systems became more commonplace.

And safer.

SCENES FROM THE EAST OHIO GAS COMPANY EXPLOSION

(Warning: graphic images)

All Photos Courtesy: The Cleveland Press Collection.

Aboveground tank destroyed by explosion.

Volunteers sift through rubble in search of victims.

Explosion aftermath.

Explosion site.

Car destroyed by unforgiving flames.

Homes on East 61st Street consumed by fire.

Aerial view of affected neighborhood.

Attending to a victim.

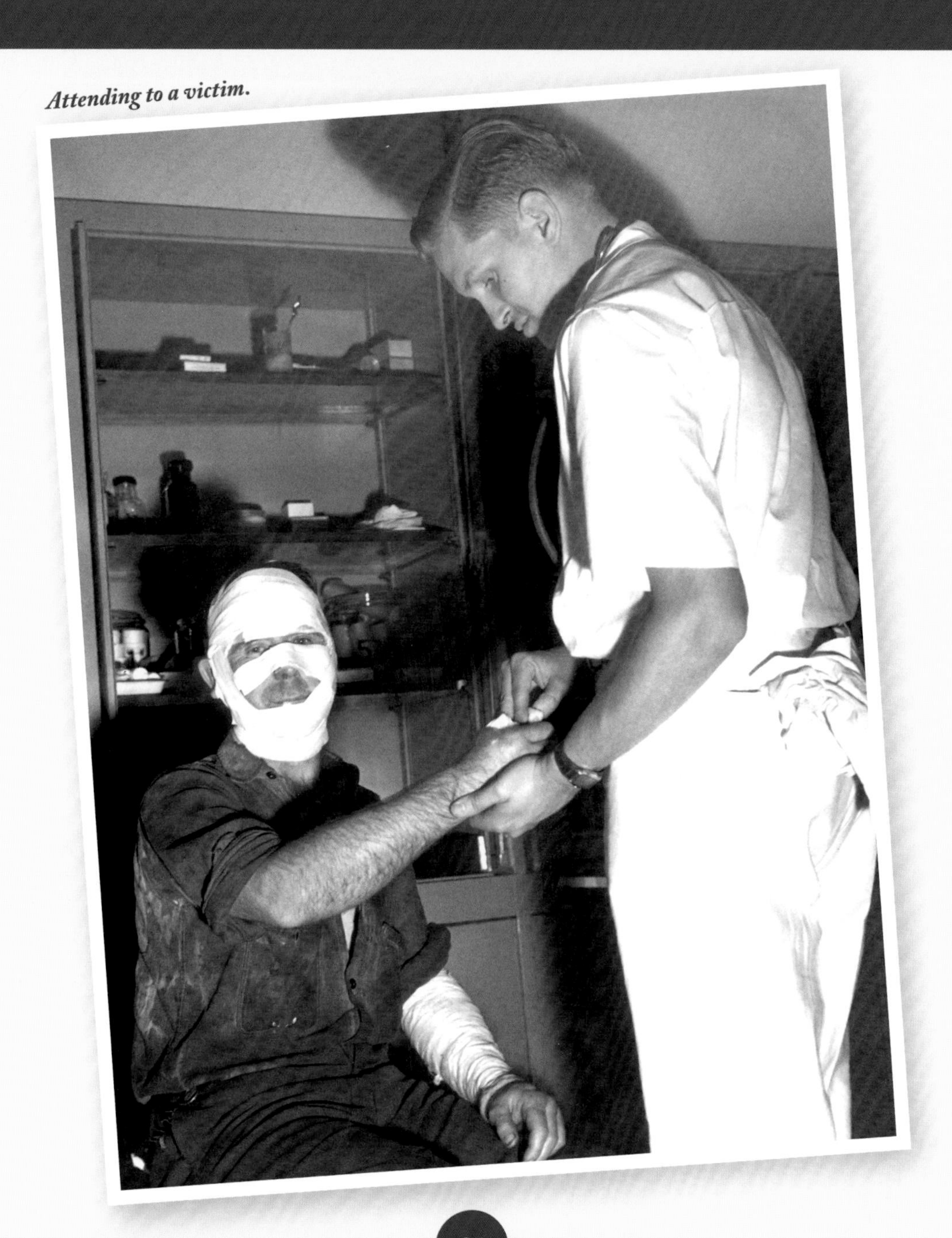

Basement of destroyed home.

Victims recover lost personal items.

Removing explosion victim.

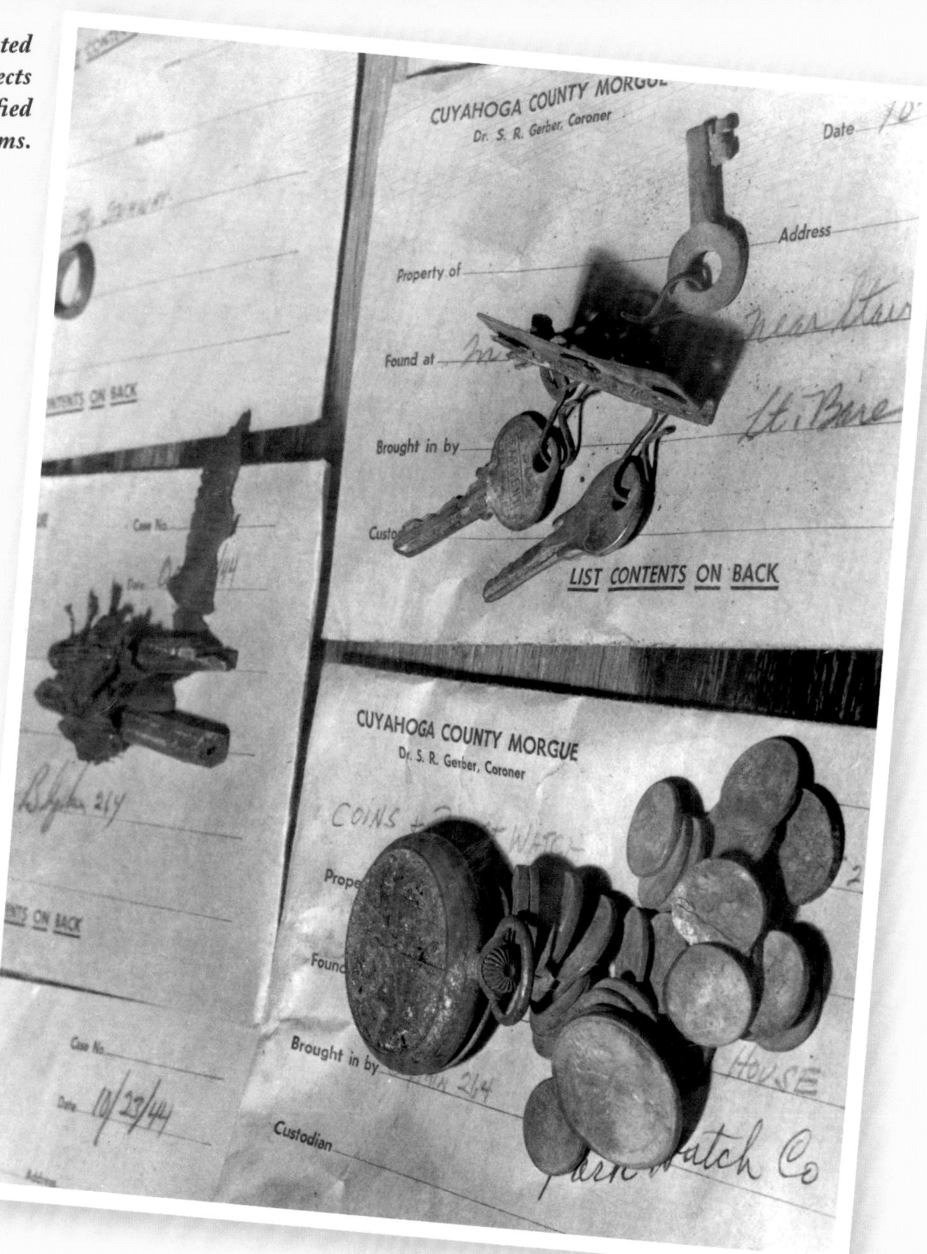

Incinerated personal effects of unidentified victims.

Rescuers remove explosion casualty.

Retrieving a casualty.

Firemen examine a fatality.

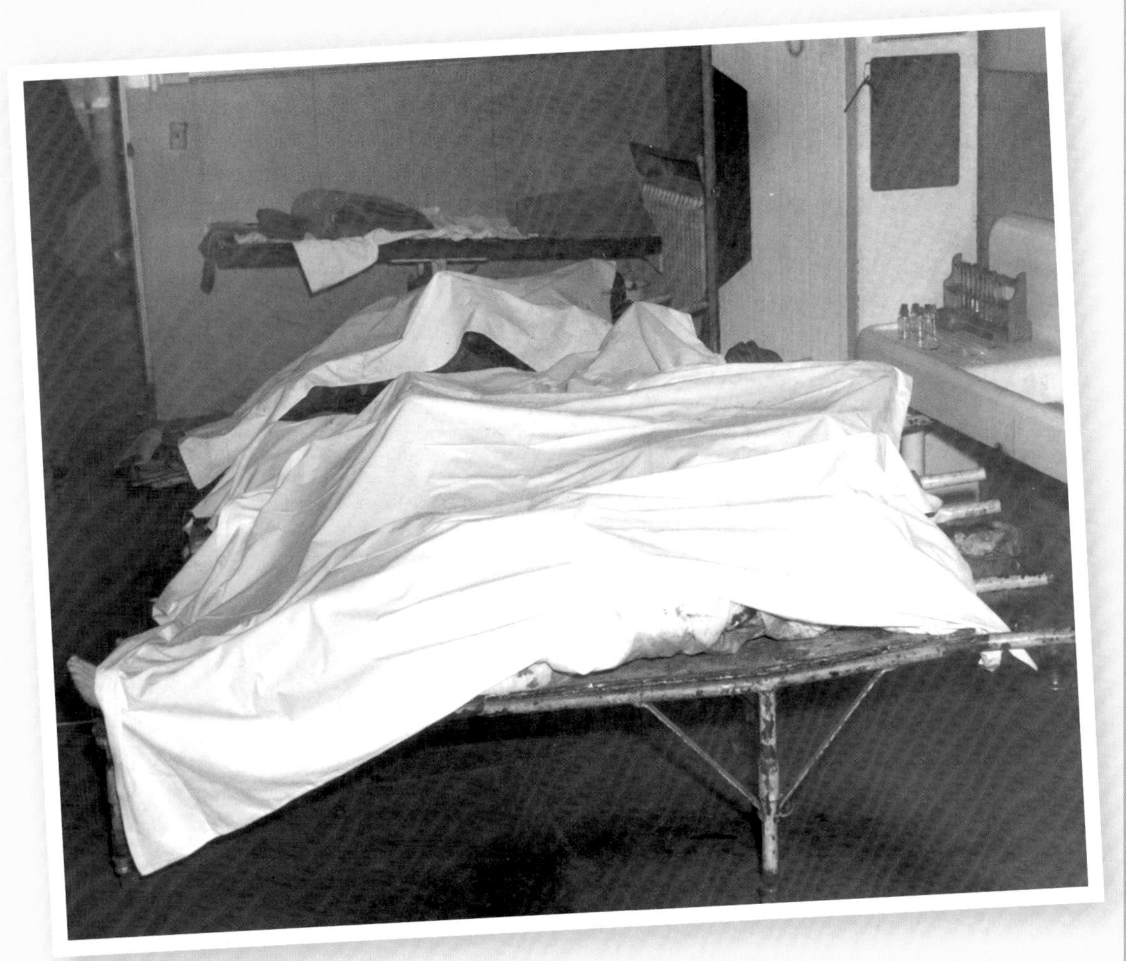

Casualties at the morgue.

Hearses line up for mass funeral.

Burial plot containing 61 unidentified victims.

Family pet discovered safe after explosion.

Ripped metal of a beer keg.

CHAPTER 6

INQUIRING MINDS WANT TO KNOW

Man stands in his own
shadow — and wonders
why it is dark.
– *ZEN PROVERB*

[SOME NAMES HAVE BEEN CHANGED TO PROTECT VICTIM PRIVACY]

BEER KEGS. We love 'em. Especially when they contain our favorite brew.

So do college kids.

And what college kid hasn't assumed his or her rite of passage by attending an outdoor keg party, the focus of which was, of course, an iced-down kegger? College is an institution encouraging students to question everything around them. So it should come as no surprise a question heard often in the educational arena is, "What would happen if . . .?"

For example: "What would happen if I threw, say, this beer can into a roaring fire?" might be typical of one experiencing an alcohol-induced shortage of judgment.

Beverage can exposed to fire.

Although seldom thought of as such, a beer can is a pressure vessel not unlike a beer keg. Not under as much pressure as a keg, a beer can is designed to accomplish a similar task: contain a frosty beverage in its entirety until accessed for one's enjoyment.

While tossing a twelve-ouncer in the campfire may be the evening's featured entertainment (*do **not** attempt*), all pressure vessels carry an element of risk and/or danger under extreme circumstances.

Many of us have seen what happens when a can is tossed into a fire. But for some, it's still amusing to watch an explosion that generates giggles among the equally intoxicated.

Getting back to our keg party: So if a can of beer can generate a chuckle or two, *what could a beer keg produce*?

Before examining the consequences of this collegiate curiosity, it might be wise to first look at what other pressure vessels do—*much smaller* pressure vessels—when exposed to heat or fire.

Those familiar with asthma inhalers know the small canisters therein contain a propellant mixture of one or more liquefied gases. What they may not know is these miniature cylinders are subject to pressure of between 50 and 80 pounds per square inch.

Lose one in a fireplace (*do **not** attempt*) and be prepared to execute a hasty exit. Like a bullet, the pressurized container will most certainly penetrate any fireplace containment screen. After ricocheting about the room, the

Exploded beverage can.

Asthma inhaler container.

capsule will, in all likelihood, make a rather deep and convincing hole in one's wall. If it doesn't first make a hole in one's person.

CO_2 capsules used in pellet guns are also pressure vessels. These containers enclose refrigerant gas exerting over 850 pounds per square inch at a temperature of 70 degrees. When the vessel is heated, CO_2 pressure increases substantially. *(One good reason not to leave a pellet gun or any type of pressurized tank in a hot car.)*

So what transpires when the ampule is heated?

Imagine sitting next to a large campfire with a diameter of, say, 12 feet and flames licking six feet toward the evening sky. You pluck one of the capsules from a box clearly marked "DO NOT INCINERATE" and pitch into the center of the fire (*do **not** attempt*).

Within seconds, the ampule detonates with a deafening explosion. Instantly, the fire is snuffed out, leaving behind nothing but hot-glowing ash and a sky full of burning embers cascading back to earth.

You read correctly: Instantaneous release of the CO_2 alone can completely extinguish flames measuring over 70 cubic feet. And all from a pressurized capsule a tad larger than a pack of chewing gum.

Now the reality: Both instances described above actually happened.

As for a beer keg being tossed into an open fire, that really happened, too.

Unfortunately, more than once.

ITEM: *It was supposed to be a festive night for a group of about 50 people who gathered for a pig roast in New Milford, Connecticut. The date was Sunday, October 22, 2006.*

There was plenty of beer. Several steel barrels used as fire pits were set aglow to warm the chilly fall evening.

Shortly after three A.M., an unidentified party-goer hoisted a quarter-keg of beer and unceremoniously dumped it into one of the burning barrels.

It didn't take long for an explosion to ensue. The loud concussion of a beer keg being instantly ripped into metal shards could be heard miles away. Windows rattled. Neighbors climbed out of their beds thinking a bomb had ignited.

Within minutes, police would arrive at the party to discover the body of 22-year-old Sean Caselli. The western Connecticut resident had been struck in the neck by a piece of metal. Another seven party attendees were taken to area hospitals with non-life-threatening burns and shrapnel-related injuries.

ITEM: *Apparently, word of the New Milford tragedy never reached Adair, Oklahoma. It was here in March 2007 a couple of residents decided to place an old beer keg into a patio fire pit.*

According to Adair Police Chief Albert McKee, "The intention was to put the keg on the fire and watch the beer spew out." But fate intervened.

"There's supposed to be a pop-off valve but they evidently decided it wasn't going to blow," McKee explained. "Well, they went inside [a house] about three minutes too early."

A resulting explosion rocked the entire Adair countryside. It not only broke out windows in three houses, metal shards cut off a four-inch tree branch "like it wasn't there." Pieces of brick from the fire pit were launched as far as a block and half away.

Chief McKee speculated the explosion "might have killed those guys if they hadn't gone back inside."

Miraculously, no one was injured.

Firefighter Trent Peper commented: "It was sure interesting, I'll say that."

ITEM: *Random acts of blissful ignorance have not been limited to the past several years. Especially when it comes to parties.*

In April 1995, an empty keg was lobbed into a bonfire at a keg party in Lake Ronkonkoma, New York.

One of the partygoers, 21-year-old Chester Vesloski, was approximately 35 feet from the fire when the keg exploded. Suffolk County Detective Sergeant Kevin Cronin stated a piece of metal severed Mr. Vesloski's arm at the elbow. He died later at the hospital.

Exploded beer keg.

None of the 10 others at the party was injured. The largest portion of the metal keg was discovered 250 feet away.

"We've had some bizarre occurrences before," officer Cronin opined, "but I've never heard of anyone being killed by a keg."

ITEM: *In 1988, at least two beer keg explosions were reported during the month of October.*

In Danbury, New Hampshire, Chris John Widebech, who had just turned 21 on Saturday, October 28, celebrated that evening with about a dozen friends. At around 12:45 A.M. on Sunday, one of the guests placed a keg into a bonfire.

Danbury police said Widebech was standing near the bonfire and was aware of the keg's location when it exploded. Reports said a piece of the aluminum keg struck Widebech in the head, killing him instantly.

Less than a week before in Albany, New York, 22-year-old Patrick Viola was killed when a keg placed in a bonfire exploded. Authorities stated Viola was struck by a shard of metal that severed his spinal cord.

While beer kegs and fire can be a lethal combination, over-pressurization can be equally deadly.

ITEM: *In 1981, a weekend party at the Sigma Phi fraternity house at the University of California at Long Beach was the scene of an explosion that killed the 25-year-old host. Police said Robert Harris died at Long Beach Community Hospital when struck by a keg.*

Reports indicated Harris and fraternity members connected the keg to a high-pressure carbon dioxide tank (designed for soft drink dispensers) pressurizing it to 75 psi. Rated at only 12 psi, the keg exploded and embedded itself in the fraternity house ceiling.

According to a police spokesman, "When they hooked up one to the other, it took off like a rocket."

Fraternity president Terry Farney said of the victim: "He was standing next to the keg. The next thing we knew there was this explosion and he hit the floor. It just happened so quick."

Harris suffered a broken arm along with internal injuries before succumbing 2½ hours later.

Perhaps one of the most unusual accidents involving a beer keg occurred in Ketchum, Idaho.

U.S. BEER KEG FAST FACTS

FAST FACT: One half keg of beer equals approximately 15½ gallons (1,984 ounces) or about 6¾ cases of 12-ounce cans or bottles.

FAST FACT: Theoretically, a full beer keg is safer than one partially depleted, given the displacement of CO_2. See the Ideal Gas Law ($PV = nRT$).

ITEM: *On January 29, 1992, Clinton Richard Doan went to a refrigerator in his garage to store a packed lunch he prepared for the next day. As he opened his refrigerator door, the bottom of a beer keg inside ruptured.*

Blaine County Chief Sheriff's deputy Gene Ramsey said Doan was killed when the keg shot upward "like a missile," striking the 35-year-old man in the head.

A faulty regulator creating extreme pressure within the keg was believed to be the cause.

It suffices to say one should not attempt to melt down a beer keg, for reasons that should now be abundantly clear. (A verity apparently lost in July 2008 on a gang of thieves from Derby, England, who learned the hard way about liquefying kegs.)

Beer and kegs share a common utility: unless used properly, the consequences can kill.

A beer keg is constructed to be useful yet harmless. But that supposition goes only as far as what one does *with* or *to* it. Therein lays potential risk.

Example: A beer keg doesn't necessarily have to be propelled by internal pressure to be dangerous.

ITEM: *On an evening in July 1990, August Sipe and Tyler Morey attended a high school reunion at a gravel pit near Rockland, Michigan.*

To enhance the evening's entertainment, both men decided to position a beer keg on some dynamite "to see what would happen." (Inquiring minds are not necessarily limited to those of inebriated college kids.)

What happened was a blast shredding the beer keg into numerous pieces of shrapnel. One piece of metal tore a hole into a nearby portable toilet. The occupant was not pleased. However, he was injured.

Messrs. Sipe and Morey were subsequently indicted and convicted of maliciously damaging and destroying personal property by means of an explosive, among other charges.

Any temptation to witness destruction of a beer keg or see the likes of the delusional Sipe and Morey should be directed to shared video sites on the Internet. Here—in the safe environs of home or office—one can view the maniacal exploits of those whose lunacy knows no bounds.

If satisfying curiosity is paramount, it's best to do it out of harm's way.

And you'll avoid losing the keg deposit fee.

CHAPTER 7

IN THE STILL OF THE NIGHT

The rest of the world is
three drinks behind.
– *Humphrey Bogart,*
American actor, 1899–1957

[SOME NAMES HAVE BEEN CHANGED TO
PROTECT VICTIM PRIVACY]

Operation of stills has always been looked upon as an enterprise of the backwoods.

Build a still, siphon its heady tonic, sell to locals. It's been a fairly lucrative formula that has appealed to numerous farmers and ne'er-do-wells looking to supplement modest earnings.

However, more and more people today are becoming involved in the distilling process to create their own personal supply in addition to generating a tidy and untaxed profit.

Any way one describes it, bootlegging is illegal. And what some of these self-styled "Hillbilly Pop" connoisseurs tend to underestimate is that this process can be very dangerous.

WANTED

information from **YOU** the taxpayer on the locations of

BOOTLEG STILLS

Moonshine stills in your locality like that pictured above, are robbing you of many thousands of dollars in Federal and State liquor taxes. Help your Government by reporting them, by mail or phone, to

ALCOHOL AND TOBACCO TAX DIVISION, INTERNAL REVENUE SERVICE

Alcohol and Tobacco Tax
Post Office Box 224
Bluefield, West Virginia Phone 9312

All communications held strictly confidential

Form 1793 (8-54)

U. S. TREASURY DEPARTMENT, INTERNAL REVENUE SERVICE

16—70765-1 GPO

The term "moonshine" was first coined in Europe in the 1700s. The reason: Moonshiners worked by the light of the moon, an obvious reference to illegal pursuits carried out during the dark of night.

"Bootlegger" emanated from colonial America and referred to bottles of spirits concealed in the boots of colonials who traded their precious home-fermented distillate to Native Americans (much to the chagrin of more-puritanical colonists).

Bootleggers and moonshiners are often portrayed in the cinema as harmless rogues providing comic relief to a story line. In reality, however, employing a still to extract your own "hooch" is a very serious and ominous process. For a number of reasons:

- Any procedure involving production of steam can be dangerous.
- Any process involving a pressure container can be dangerous.

Federal agents pose after shutting down an illegal still. Agents were called "revenuers" because moonshiners refused to pay the government liquor taxes (i.e., revenue).

- Exposing alcohol to fire is tantamount to boiling gasoline.
- In addition to an explosion that can kill or maim, incidents involving stills unleash pockets of fire not easily extinguished.
- While making moonshine may seem elementary, lack of knowledge (i.e., human error) will likely provoke an unplanned adventure.

Of course, consumption of "white lightning" is perhaps the most hazardous risk to body and soul. Alcohol distilled at home can vary from 120 to 192 proof (60 to 96 percent pure alcohol).

Still explosions over the decades have been many. However, an exact number will never be known due to the secretive, criminal nature of distilling one's own spirits.

ITEM: *Police in Chicago blamed an exploding still for curtailing the lives of two west side families in April 1927. Killed were a storekeeper, his wife, and two teenage children. Other fatalities included a tailor, his wife, and two sons.*

The families lived in back of a row of shops located in a one-story brick building housing four separate businesses. All the walls in the structure

collapsed as each of the victims lay sleeping in their beds. The 17-year-old daughter of the store-keeper was miraculously discovered alive in a pile of rubble several hours after the explosion.

Owner of the grocery store, who was not in the building at the time of the blast, was alleged to be a known bootlegger. Police reported the still was found amongst the debris.

ITEM: *The* Titusville Herald *reported in its July 10, 1927, issue: "A bushel of scorched bones and a bit of seared flesh was all that could be found today when eight prisoners from the Berks County jail combed the ashes of the little log cabin home near Bernville [Pennsylvania] . . . where Mrs. Katherine Fair and her six children burned to death Saturday morning. Some of the bones were recognized as those of Mrs. Fair, but none of the children could be identified."*

Police said there were "unmistakable signs" of a still in the house even though Mrs. Fair's seriously injured husband, Mark Fair, denied the illegal device caused the blast. Another occupant of the cabin at the time of the accident, Angelo Consulio, was not seriously injured.

According to Fair, Consulio rented the cabin's basement from him several months earlier. Fair admitted he did some hauling for Consulio to Philadelphia and nearby cities but never inquired about the truck's contents. Nor did he check the basement to learn what was being manufactured.

Fair was under the impression Consulio was making perfume and hair tonic.

Explosions during the twentieth century were common. Yet most involved "revenuers" destroying moonshine equipment with dynamite. While that may have in and of itself prevented a significant number of potential accidents, it in no way lessened the danger to those illicit vendors lucky—or unlucky—enough to escape detection. After all, distilling alcohol is not a sophisticated endeavor.

Materials required: metal pot (car or truck radiator may have sufficed in the old days but is **not** recommended); lid with small opening; coiled tube (preferably copper) fitted to hole in lid; and jar positioned at opposite end of tube.

Ingredients: corn meal, sugar, water, yeast, malt.

Instructions: Bring mash to light boil in pot, allowing alcohol vapor to escape through coiled tube. Collect alcohol condensate in jar. Boil condensate repeatedly as necessary to achieve desired proof.

What can go wrong? Although most stills are not high-pressure operations (many pressure cookers have a working pressure setting of 15 psi), clogging the tube or lid opening can provoke explosion, as can exposing high-proof

25
DOLLARS PER YEAR
Stub for
Special Tax Stamp.
No G97333
Issued to Collector.
To
Retail Liquor Dealer
at
on the day of 187
for period commencing
1st 187
ending
April 30th 1880.
$
AMOUNT OF TAX

Coupon for RETAIL LIQUOR DEALERS SPECIAL TAX for May 1879.
Coupon for RETAIL LIQUOR DEALERS SPECIAL TAX for June 1879.
Coupon for RETAIL LIQUOR DEALERS SPECIAL TAX for July 1879.
Coupon for RETAIL LIQUOR DEALERS SPECIAL TAX for Aug. 1879.
Coupon for RETAIL LIQUOR DEALERS SPECIAL TAX for Sep. 1879.
Coupon for RETAIL LIQUOR DEALERS SPECIAL TAX for Oct. 1879.
Coupon for RETAIL LIQUOR DEALERS SPECIAL TAX for Nov. 1879.
Coupon for RETAIL LIQUOR DEALERS SPECIAL TAX for Dec. 1879.
Coupon for RETAIL LIQUOR DEALERS SPECIAL TAX for Jan. 1880.
Coupon for RETAIL LIQUOR DEALERS SPECIAL TAX for Feb. 1880.
Coupon for RETAIL LIQUOR DEALERS SPECIAL TAX for March 1880.
Coupon for RETAIL LIQUOR DEALERS SPECIAL TAX for April 1880.

$
No G97333
United States
Stamp for Special Tax
Internal Revenue
1879
Received from the sum of
100 Dollars for Special Tax
on the Business of Retail Liquor Dealer
to be carried on at
State of for the period represented by the Coupon or Coupons
hereto attached. Dated at
187
Collector. Dist.
State of
25
DOLLARS PER YEAR
SEVERE PENALTIES are imposed for neglect or refusal to place and keep this Stamp conspicuously in your establishment or place of business.

Coupons such as those above (circa 1870) had to be purchased yearly by retail liquor dealers in order to sell alcohol for consumption. For moonshiners, paying the "special tax" was never a consideration.

alcohol to an open propane flame. (Some professional distillers do not permit use of flash cameras during factory tours, lest alcohol fumes be ignited.)

Typically, about five gallons of mash are required to squeeze out a gallon of 150-proof distillate.

Those who think distilling one's own bathtub gin is a product of yesteryear might want to access online auction site eBay, where more than 100 items were recently featured ranging from how-to booklets to actual still equipment. The online video-sharing site YouTube boasted over 450 videos, many of the how-to variety.

ITEM: *A September 2008 article in the Athens, Alabama,* News Courier *reported a father and son were injured when a large still exploded in their brick garage. "It was like something out of Prohibition," commented Limestone County Sheriff's Investigator Randy Burroughs. Jerome Kessler was admitted in serious condition at a Birmingham hospital while son Jay underwent treatment for eye injuries. "I thought it was probably a meth lab," said Burroughs of the one-A.M. Sunday explosion.*

"When I got to the garage, I was looking the still over and I couldn't figure out where the relief valve was," he continued. "I could see they forgot to take the plug out of the lid. Instead of cutting off the heat and letting it cool off so they could remove the plug, Jay told me they were trying to knock the plug off to release the pressure so they could take out the plug. That's when it exploded."

The Courier *said molten vapor burned the elder Kessler's chest and back. Investigators found three 50-gallon garbage cans in a cellar under the garage filled with wine and mash.*

Burroughs observed the illegal concoction was not run-of-the-mill wine: ". . . it's nearly pure grain alcohol, which is 180 proof!"

CHAPTER 8

REMEMBERING THE ALAMO: The Roundhouse Explosion of 1912

The greatest of faults is to be conscious of none.
– *Thomas Carlyle, Scottish philosopher, 1795–1881*

On the morning of March 18, 1912, a slight breeze blew through the east side of San Antonio.

One of numerous railroad communities dotting the Texas landscape, the Alamo city was a key Lone Star State terminal, where many of the male population worked on regional rail systems. On Austin Street, the Southern Pacific Roundhouse was a focal point of railroad activity.

And so it was on this balmy morn that the ladies of the community commenced their daily routines.

After getting her husband—a blacksmith—off to work, Mrs. Zysko awakened her two children in preparation for the day ahead. Mrs. McCall leisurely strolled across her yard as Mrs. Peters stood in hers admiring the southern Texas morning. Nearby, 82-year-old Mrs. Gillis went about her normal household chores, as did Mrs. Stephens, her neighbor. Mrs. Howard left her house to retrieve the daily mail.

Several blocks away, roundhouse workers toiled diligently to return a modern 10-wheel passenger engine to service on the Galveston, Harrisburg, and San Antonio railroad line. Having been involved in a wreck the previous December, locomotive No. 704's repairs were complete. Shop painter Jose E. Fuentes stood aside the driving rod where he applied finishing touches to an insignia.

Machinist helper Robert Mantiel positioned himself next to the engine and looked on as yeoman engineer Walter Jourdan tinkered about the engine's exterior, checking fittings and making last-minute adjustments.

Nearby, blacksmith and machinist shops were alive with activity as scores of workers performed assigned duties. There were cooper P.J. Stoudt, pipefitter helper Archie Price, blacksmith foreman Henry Mansker, and his son James.

Many in the shops were workers employed to replace boiler repairmen, copper fitters, and other craftsmen who went on strike in the fall of 1911. "Strikebreakers," as they were called, came from the northeast as well as cities across the Texas Plains. And they all shared something in common: Each was new to the rail yard and San Antonio.

Some were so new their names had not yet been recorded by the railroad. Labor issues became so contentious that assistant roundhouse foreman James Valentine began carrying a pistol.

But the Galveston, Harrisburg, and San Antonio line had a railroad to run. The priority of that March morning in Texas was to return No. 704 to service. That day, initial firing of the boiler generated some concern about the engine's pressure valves. The pressure lowered, the boiler was re-fired and all appeared normal.

Mechanics of a steam age locomotive are relatively simple: A cylindrical water boiler sits atop a rail carriage. A firebox within the boiler burns coal or wood to generate steam pressure to drive pistons. The pistons activate driving rods that propel carriage wheels.

The amount of pressure required must be enough to tow thousands of tons of railroad cars, as well as accompanying freight and passengers.

At 8:55 A.M., as No. 704 sat idling, superheated steam suddenly ripped open the boiler's iron casing with a pressure that separated one-ton iron wheels from axles. As the cylinder released from the carriage, it became vertically airborne along with hundreds of metal parts from the engine.

Concussion from the instant release of energy collapsed roundhouse walls and displaced all within its path, including heavy equipment and people. Everything within close proximity of the engine instantly disintegrated. Outward pressure propelled shards of metal, wood, and body parts beyond the railroad yard and into surrounding neighborhoods. For the next several moments, no one in the community was protected from what would become known as *the greatest locomotive boiler explosion in railway history.* Massive trees were uprooted, windows shattered, and the sky rained terror from which there was no escape.

Impact of the explosion and resulting earth tremor froze east San Antonio residents where they stood. Some thought they were experiencing an earthquake.

Seven blocks from the roundhouse, Mrs. Gillis was fatally injured when a front portion of the locomotive smashed through her roof and destroyed three rooms before finally impaling her in the ground floor of her house. A wooden one-inch splinter pierced the elderly woman's skull.

Mrs. Howard was returning to her house when a large portion of the boiler's crown plate fell from the sky and destroyed the back rooms of her home. The several-hundred-pound metal shard also created a hole in the ground floor.

Homeowner W.H. Witer felt the impact of an object crashing through his roof. He would later discover a human arm stuck in a rafter.

Mrs. Stephens sustained severe injuries when a segment of the engine's water jacket pulverized an appreciable portion of her home.

Mrs. McCall was stunned to observe a disfigured torso drop into her yard. Several moments later, the San Antonio woman's horror was compounded when an auxiliary air tank fell directly into her path.

As she stood by her gate, Mrs. Peters was hit with an airborne iron bolt. The fastener knocked her unconscious and caused critical internal injuries.

Back at the accident scene, a black plume of smoke snaked from where the roundhouse once stood. Anything and everything within a 100-foot radius of No. 704 was no longer of this Earth. About 30 percent of the heavy equipment and machinery was either dispatched from the roundhouse or existed as mangled wreckage. The *New York Times* reported: "Boiler tubes are scattered all over the grounds, in some places 1,000 feet from the scene of the explosion."

That which remained was covered with black oil, the result of an oil tender behind No. 704 being launched on its tracks against another locomotive 150 feet away. It took only seconds for the oil to ignite, further complicating an already catastrophic situation.

As fires raged, military corpsmen and civilians commenced a search and rescue of survivors. They were surprised to find several who were miraculously alive.

James Mansker had been knocked unconscious during the explosion. Regaining his bearings, he rose from the wreckage in search of his father. Finding the elder Mansker dead in what had been the blacksmith shop, James lifted the old man and carried him home.

P.J. Stoudt and foreman T.A. Williams were saved when concussion from the blast blew them beneath an industrial workbench. Like several others who survived, the men were blown under twisted steel that shielded them from a collapsing roof.

Not so fortunate was Walter Jourdan, whose body was found not too far away. The thrashed, contorted remains made identification of the engineer's body a difficult and time-consuming obligation. The human forms of Robert Montiel and Jose Fuentes were no longer. It took the removal of Archie Price's body to one of the morgues before identification could be established. James Valentine's body was recognizable only by the pistol he had with him.

Almost every one of the discovered bodies was disarticulated. All were covered with thick black oil from the destroyed tender. And all displayed small pieces of wood and coal that had penetrated their flesh.

Three other locomotives in the now-destroyed roundhouse were also damaged. A steam whistle on one had been bent open releasing a plaintive, one-note calliope-like sound that droned for two hours before its pressure had been—to the relief of rescue workers—released.

As horse-drawn ambulances began removing bodies, two-man teams of YMCA workers headed to the surrounding neighborhoods. The teams carefully culled San Antonio's east side with baskets covered by a white cloth, the purpose of which was to collect body parts and human tissue scattered for blocks throughout the small community. They also carried long poles used to disengage remains that had fallen onto power lines and into trees.

Word of the disaster spread quickly. The local streetcar company added extra cars to transport more than 50,000 curious onlookers to the macabre scene.

Attempting to find her husband, the wife of Carl Zysko arrived at the yard only to learn of the blacksmith's death. One of the recent arrivals to San Antonio who spoke no English, Veronike Zysko—hugging a small child—sobbed relentlessly.

In all, 26 men—mostly "strikebreakers"—died and 40 were injured. Fourteen in the roundhouse were instantly dismembered. Collected remains indicated at least 10 others had perished but never counted. Published reports put the number of dead workers and area residents at between 40 and 50; a complete tally of residents killed and injured was never taken. Monetary losses ranged from $250,000 to $750,000 (about $5.8 million and $17 million respectively in 2012 dollars). These losses did not include damage to the community.

On March 25, the San Antonio City Council considered a proposed ordinance creating a boiler inspector's position and requiring the qualification and certification of all persons servicing stationary boilers and locomotives within the jurisdiction. The second of three required readings of the regulation took place two weeks later. The third reading, however, never occurred. Nor did the proposal become an ordinance.

Investigators would later speculate the No. 704 boiler explosion resulted from one of several scenarios.

Sabotage on the part of strikers seemed to be a real possibility given the absence of the locomotive's fireman and repair supervisor who should have been present during the boiler-firing process. A pressure relief valve recovered following the accident revealed it had been screwed in place.

Another theory advanced about a week after the tragedy involved nitroglycerin. According to the March 18, 1912, *New York Times*, "The theory of nitroglycerin appears supported to a certain extent by the fact that the locomotive which blew up had passed a thorough inspection test only two days ago."

Human error was also examined. Initial firing of the boiler revealed pressure problems. The second firing was undertaken after the crew was ordered to prepare the locomotive for service later that day.

Were shortcuts taken to meet the company's objective?

Were strike replacement workers fully knowledgeable of the firing procedure? Similar explosions occur when a hot boiler has little or no water and water is subsequently introduced.

One thing is certain: The boiler did suffer mechanical failure. Was this a consequence of the December accident? Given the limited familiarity of metallurgical technology in 1912, we will never know for certain.

Railroading was a very dangerous profession during this era. Between 1890 and 1917, more than 72,000 employees were killed with another 2 million injured. Incredibly, another 158,000 lives were shortened in repair shops and roundhouses.

As for conclusions of the investigation, perhaps the most detailed explanation came from Texas Labor Statistics Commissioner J. A. Starling, who told the *San Antonio Express*:

> Speaking from forty years' experience in this line of work, I am convinced beyond doubt that the occurrence was caused by tremendous boiler pressure. The engine was one of the biggest in the country, and there was not a piece of the boiler as big as that [referring to a small bookcase] to be found after the explosion. Of course there was no telling just what the pressure was, as the steam gauge was wrecked, or whether there was the proper supply of water, but there is no question that boiler pressure was responsible for this terrible accident.

Captain Guy Carleton, assistant chief of the Bureau of Explosives for the American Railway Association, made but one public observation: *The main force of the explosion was steam.*

He provided no comment regarding the cause.

Site of the roundhouse explosion is located today at the Southern Pacific yard near Interstate 35 and Austin Street in San Antonio.

SCENES FROM THE ROUNDHOUSE EXPLOSION OF 1912

ABRIDGED LOCOMOTIVE BOILER EXPLOSION TIMELINE

NOTE: No comprehensive records were kept by either the government or industry during many of the following years:

1815

Philadelphia, Co. Durham, Northern England. Experimental locomotive *Brunton's Mechanical Traveller*, (four wheels pushed by mechanical feet) exploded (16 dead, mainly sightseers). Acknowledged as earliest recorded locomotive explosion.

1828

Stockton on Tees and Darlington, Northern England. Boiler on Locomotive No. 1 exploded (1 dead, the driver).

1831

The Best Friend, Charleston, South Carolina. Exploded after a fireman tied down the safety valve (1 dead, the fireman). 17 June.

1856

Fort Washington, Pennsylvania. Two passenger trains collide, one carrying 1,500 Sunday school children to a picnic. The boiler on the other passenger train exploded. Although no deaths resulted directly from the boiler explosion, 50 to 60 children died in the crash. One of the conductors committed suicide the same day but was later absolved of any responsibility. 17 July

1865

United Kingdom. Ten explosions during a single year, nine resulting from corroded boiler barrels.

1881

New South Wales, Australia. E17 class locomotive exploded. 21 December.

1892

North East Dundas Tramway, Tasmania. 0-4-2 locomotive exploded. 18 June.

1894

Tucuman, Argentina Republic. (8 dead). 24 July.

1896

West Texas. The Crash at Crush occurred when a promoter staged a deliberate train wreck. Boilers on the two vehicles exploded, launching shrapnel toward a crowd of 30,000 to 40,000 paying spectators (3 dead). 15 September.

1904

Morristown, Tennessee. Two Southern Railway passenger trains collided head-on. Impact knocked exploding boilers off both engines, causing one engine to catapult onto three wooden coaches of the other locomotive. 24 September.

1910

Cardiff, Wales. Fitter incorrectly assembled locomotive's safety valves. Boiler exploded in shed (3 dead). 21 April

1912

San Antonio, Texas. Boiler failed on a Southern Pacific locomotive at a Roundhouse. One victim was killed in her home seven blocks away when a boiler crashed through the roof (26 total dead). 18 March.

1921

Buxton, UK. As a result of safety valves being too tight, the boiler was estimated to have reached 600 psi before exploding. Design pressure—200 psi. 11 November.

1925

Rockport, New Jersey. Seven-car Lackawanna Railroad passenger train traveling to Hoboken, New Jersey, derailed during rainstorm. Subsequent engine boiler explosion scalded passengers (51 dead). 16 June.

1943/1944

United Kingdom. Three American-built locomotives exploded in service due to operator unfamiliarity with water gauges.

1948

The last boiler explosion on the Union Pacific Railroad occurred aboard a 9000 class 4-12-2 (3 dead). 20 October.

1962

Bletchley, UK. Last boiler explosion on British Railways. Lack of water caused firebox to collapse. 24 January.

1977

Bitterfeld, East Germany. Lack of water on a Class 01 steam engine caused explosion (9 dead, 45 injured). 27 November.

CHAPTER 9

DID YOU KNOW?

The question isn't who is going to let me; it's who is going to stop me.
– Ayn Rand, Russian-American philosopher and novelist, 1905–1982

- **The first full-sized airship was steam-powered.**
  Employing a steam engine for power, French inventor Henri Giffard designed and flew the first dirigible. On September 24, 1852, he navigated the 3-horsepower craft approximately 15 miles west from Paris to the small community of Trappes at a speed of about six miles per hour. The unique aircraft consisted of a gas-filled, football-shaped balloon surrounded by a net. A gondola suspended from a pole connected to the net carried the passenger along with a 100-pound boiler and 250-pound engine.

- **The first airplane was steam-driven.**
  The steam-powered Aerial Steam Carriage was designed by English aviation engineer William Samuel Henson and British lace-making engineer John Stringfellow. An unmanned 10-foot model was flown in 1848 and was the world's first heavier-than-air powered flight.

Steam-powered Aerial Steam Carriage.

- **The first helicopters were steam-driven.** A model steam vertical lift aircraft created by Englishman William Henry Phillips attained flight in 1842 and was the first to employ a small power plant as opposed to stored energy mechanisms (i.e., wind-up devices). Featuring a rotor diameter of 15–18 inches, the two-pound craft achieved lift by way of cranked tubes injected with steam from a miniature boiler. Some credit Frenchman Ponton d'Amecourt with pioneering the first vertical lift aircraft in the early 1860s. He called his model steam-powered models helicopteres, derived from the Greek adjective "elikoeioas" (spiral or winding), and noun "pteron"(feather or wing). Interestingly, Phillips changed priorities in 1849 when he successfully fashioned his boiler to power one of history's earliest fire extinguishers.

- **The first power boat was steam-driven.** While it is believed the very first paddle boat was pedal-powered and constructed by French inventor Denis Papin in 1707, it was later equipped with a steam pump and first navigated down the Fulda River in Germany. With his boat believed to be far ahead of its time, Papin could not interest investors to finance his project.

- **The first locomotive was steam-driven.**
  Industrial Revolution engineer Richard Trevithick constructed the first full-scale working steam-powered vehicle to replace horse-drawn carts on tramways. It was demonstrated in England on February 21, 1804, while hauling 10 tons of iron, 70 men and five wagons a distance of nine miles over a two-hour time span.

- **The first motorcycle was steam-powered.**
  American inventor Sylvester Roper created a steam-driven and the first pedal-less motorcycle in the late 1860s. He constructed his second steam-driven motorcycle in 1895.

- **The first U.S. vehicle fatality involved a steam-driven motorcycle.**
  Sylvester Roper died of a heart attack on his second motorcycle in 1896 while serving as a pacesetter for a bicycle race. It was believed the heart attack was provoked by "the excitement attendant upon the remarkable speed developed at the exhibition."

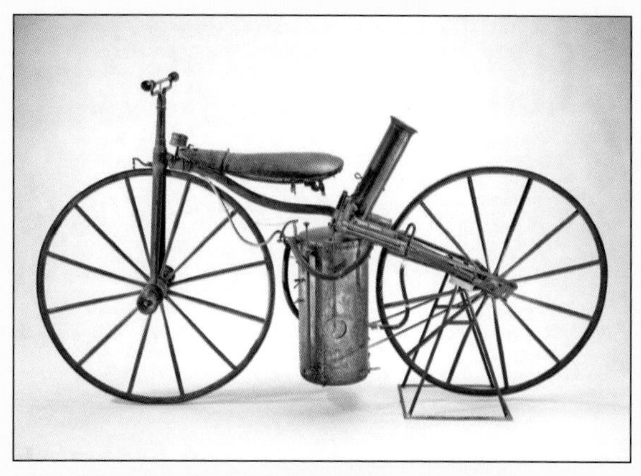

Sylvester Roper's first motorcycle now resides permanently at the Smithsonian.

- **The first recorded locomotive boiler explosion occurred in 1815.**
  The experimental railway locomotive Brunton's Mechanical Traveller exploded on July 31, 1815, in County Durham, England. The incident resulted in the deaths of sixteen people, a majority of whom were curious onlookers. Also called the Steam Horse, the newfangled engine moved on four wheels pushed by mechanical feet. Because the locomotive ran on an industrial wagon way as opposed to a railway, the accident is seldom recognized as the earliest boiler accident.

- **Steamboat boiler explosions resulted in more than 4,000 estimated fatalities from 1816 to the 1850s.**
  The number of vessels destroyed exceeded 500.

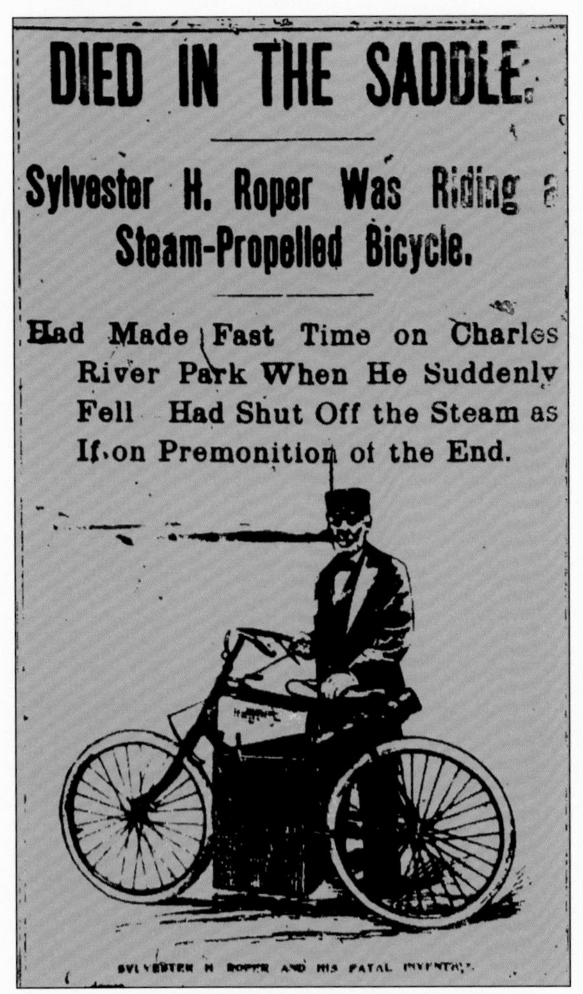
DIED IN THE SADDLE.

Sylvester H. Roper Was Riding a Steam-Propelled Bicycle.

Had Made Fast Time on Charles River Park When He Suddenly Fell Had Shut Off the Steam as If on Premonition of the End.

SYLVESTER H. ROPER AND HIS FATAL INVENTION.

Announcement of Sylvester Roper's death in the Boston Daily News, *June 2, 1896.*

- **In the ten years between 1895 and 1905, it was estimated over 7,600 individuals—an average of two per day — were killed by boiler explosions in the United States.**
  Between 1885 and 1895, over 200 boiler explosions were reported per year. The following decade saw more than 3,600 such explosions or approximately one each day.

- **Water converted to steam instantly expands 1,600 times in volume.**
  One thousand gallons of water vaporizes to over 1.5 million gallons of steam in less than one second!

- **A water heater explosion can eject the tank upward through several floors of a typical home.**
  Rupture of a typical 30-gallon home hot water tank generates enough force to send a 2,500-pound car to a height of nearly 125 feet (equivalent of a 14-story building) with a liftoff velocity of 85 miles per hour. It has been documented a similar tank has been launched from the basement through the roof of a three-story farmhouse.

- **The explosion capacity of a 30-gallon water heater is comparable to the explosive potential of nitroglycerin.**
  The instant release of over 300,000+ foot-pounds of energy is equal to the explosive force of 0.16 pounds of nitroglycerin.

- **One of the most prominent symbols in old Yankee Stadium was actually a boiler stack.**
  The 138-foot baseball bat outside the main stadium entrance at Gate 4 covered a boiler vent. Adorned with Louisville Slugger Logo and Babe Ruth's signature, the stack was fitted to resemble a bat complete with a knob at the top and a taped handle. For

Yankee fans, it was common to meet each other, "at the bat."

- **Sudden release of pressure converts a compressed gas cylinder into a missile-like projectile that can penetrate concrete block walls.**
  Gas cylinders retain extremely high pressure, up to 2,500 pounds per square inch (psi).

- **A tractor-trailer tire explosion can be lethal.**
  Tests have revealed explosion debris can penetrate a car window with enough force to decapitate the driver and/or passengers.

- **Metal debris from industrial boiler accidents can be ejected as far as a quarter of a mile.**

- **The smallest working stationary steam engine measures just over ½ inch.**
  With dimensions of 0.267 inches high and 0.639 inches long, the miniaturized engine is constructed of a cylindrical body of brass with built-in crankshaft bearings, piston, connecting rod, and valve mechanism. Creator Iqbal Ahmed of Nagpur, India, says the engine can run for two minutes on 10 cc of water, powered by a separate boiler (less than two inches in height) and heated with an alcohol burner. Total steam engine weight: 0.06 ounces.

- **Water boils nine degrees lower in mile-high Denver than at sea level.**
  At higher altitudes, water boils at lower temperatures than at sea level. Mountain climbers often use pressure cookers to compensate for lower atmospheric pressures at higher elevations.

- **Steam hybrids recovering waste heat could reduce fuel consumption up to 31.7%.**

- **The human body is technically a pressure vessel.**
  Although basically containing water mass and a skeletal support system, the human

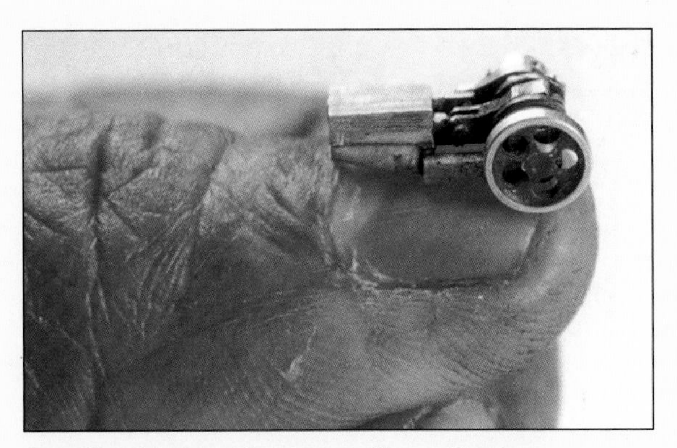

Smallest Working Stationary Steam Engine.

body with its network of blood vessels can be considered a pressure vessel. It can withstand 50 psi blast pressure (i.e., sudden impact). Under sustained pressure that is gradually increased, the body can withstand as much as 400 psi.

- **One of Houdini's greatest stunts involved escaping from a galvanized iron riveted boiler.**
  Houdini was riveted into a large, hot water boiler onstage by actual local boilermakers. To secure the master magician in place, the entrance door was threaded with a hard steel bar padlocked at both ends. The boiler with Houdini was then placed inside a cabinet. The audience fretted for the performer's safety as the band played a lively (and loud) medley selection. Thirty minutes later, a winded and sweat-laden Houdini would emerge from the cabinet to rousing applause. How'd he do it? Let's just say the **hard** steel bar was mysteriously replaced with a **soft** steel bar and the noisy process of sawing through it was effectively drowned out by an unwitting yet complicit orchestra.

- **Although not constructed as such, caskets are also pressure vessels.**
  Funeral home professionals call it Exploding Casket Syndrome when methane gas is created in the casket during body decomposition. A phenomenon primarily occurring years ago in above-ground mausoleums, the outcome can blow off the lids of caskets as well as dislodge marble door panels on crypts. Casket manufacturers today include safety relief devices called "burpers" (gaskets allowing the release of accumulated gases) on sealer caskets.

- **It pays to recycle. Even in death.**
  For years, the Swedish west coast town of Halmstad had been searching for a way to curb the amount of smoke emitted from the local crematorium. A community with a proud environmental conscience, the fine citizens of Halmstad in 2008 came up with a rather profound approach not only to dispatch the smoke, but dispatch it in a way to extract heat heretofore escaping into the surrounding fjords. Public discourse led to routing the heat from the town crematorium to local homes. "It was when we were discussing all these environmental issues that we started thinking about the energy that is used in the cremations and realized

Harry Houdini.

that instead of all that heat just going up into the air, we could make good use of it somehow," explains Halmstad cemetery director Lennart Andersson. "It was just rising into the skies for nothing."

- **A renowned American Inventor in 1911 predicted the imminent death of the steam engine.**
  Thomas Edison told the *Miami Metropolis* on June 23: "… the steam engine is emitting its last gasps. A century hence it will be as remote as antiquity as the lumbering coach of Tudor days." Among the incandescent icon's other visions: "Gold has even now but a few years to live. The day is near when bars of it will be as common and as cheap as bars of iron or blocks of steel."

- **U.S. military considered use of steam-powered warplanes in the early 1930s as a way to avoid detection by sound.**
  H.J. Fitzgerald wrote in the July 1933 *Popular Science Monthly:* "Because above 1000 feet, steam-driven planes would be as silent as soaring birds, they would have particular value in military work. Noiseless warplanes have long been sought. But muffling gasoline engines reduces their power to such an extent that the plan is impractical. The new power plant, silent by nature, would permit long-distance raids above the clouds by ghost ships giving off no telltale drone of motors to warn the enemy or to aid in directing anti-aircraft fire."

- **The Apollo XIII scheduled moon landing was aborted after an oxygen tank exploded.**

- **Steam engines were critical to helping farmers feed America in the late 1800s and early 1900s.**
  Among the engines' many uses: threshing wheat and other small grains, running saw mills, steaming tobacco, shelling corn, cutting silage, running husker-shredders, hauling, extracting tree stumps, and providing heat.

- **Of the thousands upon thousands who have died as a result of pressure equipment accidents, it is believed at least one death has been associated with paranormal phenomena.**

The year was 1951 and the location was the KiMo Theater in Albuquerque, New Mexico. Frightened by what he witnessed on the theater's movie screen, six-year-old Bobby Darnall left his seat in the balcony and ran down to the lobby. As Bobby reached his destination, an eight-gallon electric water heater exploded under the lobby floor, thus resulting in Bobby's death and injury to nine others. Not long thereafter, theater patrons and staff reported sightings of an apparition appearing to be a small boy wearing blue jeans and a striped shirt playing on the lobby staircase. The sightings were accompanied by events unexplained to this day. To appease the reported apparition, theater staff later established a shrine to Bobby consisting of gifts along a wall behind the stage. Some believe the spirit of Bobby Darnall continues to haunt the KiMo Theater to this day.

- **Stanley Steamer steam cars from the early 1900s boasted less than 40 moving parts.**

Combustion engine cars of the same era contained more than 500 moving parts. Estimates of the number of parts in cars of today vary widely from thousands for a conventional gas engine automobile, to several hundred for an electric vehicle.

- **Fire extinguishers will not explode when exposed to fire.**

Recently manufactured extinguishers contain a release valve which actuates if the internal tank pressure is exceeded, thereby safely releasing carbon dioxide.

- **NASA did not create the space pen.**

It was actually Paul Fisher who fashioned a pressurized canister that forced thixotropic ink to the ball tip irrespective of ambient pressure. The inventor brought the pen to the space agency where it was first used by astronauts on Apollo 7 in 1968.

- **Cars could automatically cook food.**

Touring cars enjoyed significant popularity in the 1930s. Catering to the appetites of this new nomadic demographic, *Modern Mechanix* magazine touted the Automatic Food Cooker. Mounted on the back end of a touring car bumper, this steam pressure kettle was promoted as an ideal way to prepare hot meals for traveling families. Just connect it to the car exhaust pipe and quicker than one can say *bon appétit*, a tasty almost home-cooked dinner awaits. (Actually, an hour's drive was required to thoroughly cook meats and vegetables.)

Automatic Food Cooker Runs by Exhaust Heat of Car

Tourists can cook their dinners while traveling through use of this cooker heated by car's exhaust.

MEALS can literally be cooked on the run through the use of the automatic cooker shown in the photo above. The cooker is mounted on the rear bumper of the motor tourist's car and an extension from the exhaust pipe connected up with it, as shown in the insert. The cooker contains a steam pressure kettle which is heated by the hot exhaust gases. An hour's drive is quite sufficient to thoroughly cook meats and vegetables. Total weight of the unit is so slight that running qualities of the car remain quite unaffected. Motor tours are much more pleasant when one is assured of a well-prepared meal at the end of the trip.

While a seemingly unappetizing way to prepare food, the Automatic Food Cooker only used heat from the exhaust, not exhaust gases. Fumes never came in contact with food. Similar technologies are applied today to recover heat from exhaust gases and are considered ecologically smart. Indeed, the Automatic Food Cooker may have been well ahead of its time.

- **The only invention patent held by a U.S. president involved steam and pressure vessels.**

Having spent part of his early life on the river, this inventor became intimately familiar with problems synonymous with steamboats, including running aground on sand bars. Once grounded, the ship required insertion of anything floatable—barrels, loose planks, boxes—under the hull's sides to provide buoyancy. The inventor's river experience and inquisitiveness inspired him to construct an 18-inch flatboat model using inflatable chambers located on each side of the boat's hull just under the waterline. By way of a scheme involving pulleys, shafts, and ropes, either one or all four bellows would be filled with air using the boat's steam power. The objective: Raise the hull over the obstruction, thus precluding the necessity of displacing cargo to increase buoyancy. The young congressman publicly demonstrated his process by putting the model afloat in the water with several bricks sinking it to the first deck. Applying air pumps modeled like old fire bellows, the model slowly but successfully rose above the water. The legislator brought his model to Washington, D.C., where he proceeded to the U.S. Patent Office and was granted a patent two months later. Patent document in hand, the congressman's invention was never marketed or publicized. Or heard of again. Some historians theorized the weight of the boat-lifting apparatus would only add to the problem rather than provide a solution. Others suggested the patent may have been instrumental in the design of modern submarine construction and ship salvaging operations. Whatever became of the model holding so much promise? It is displayed at the Smithsonian Institution. And the inventor? He was assassinated in 1865 at the hand of John Wilkes Booth.

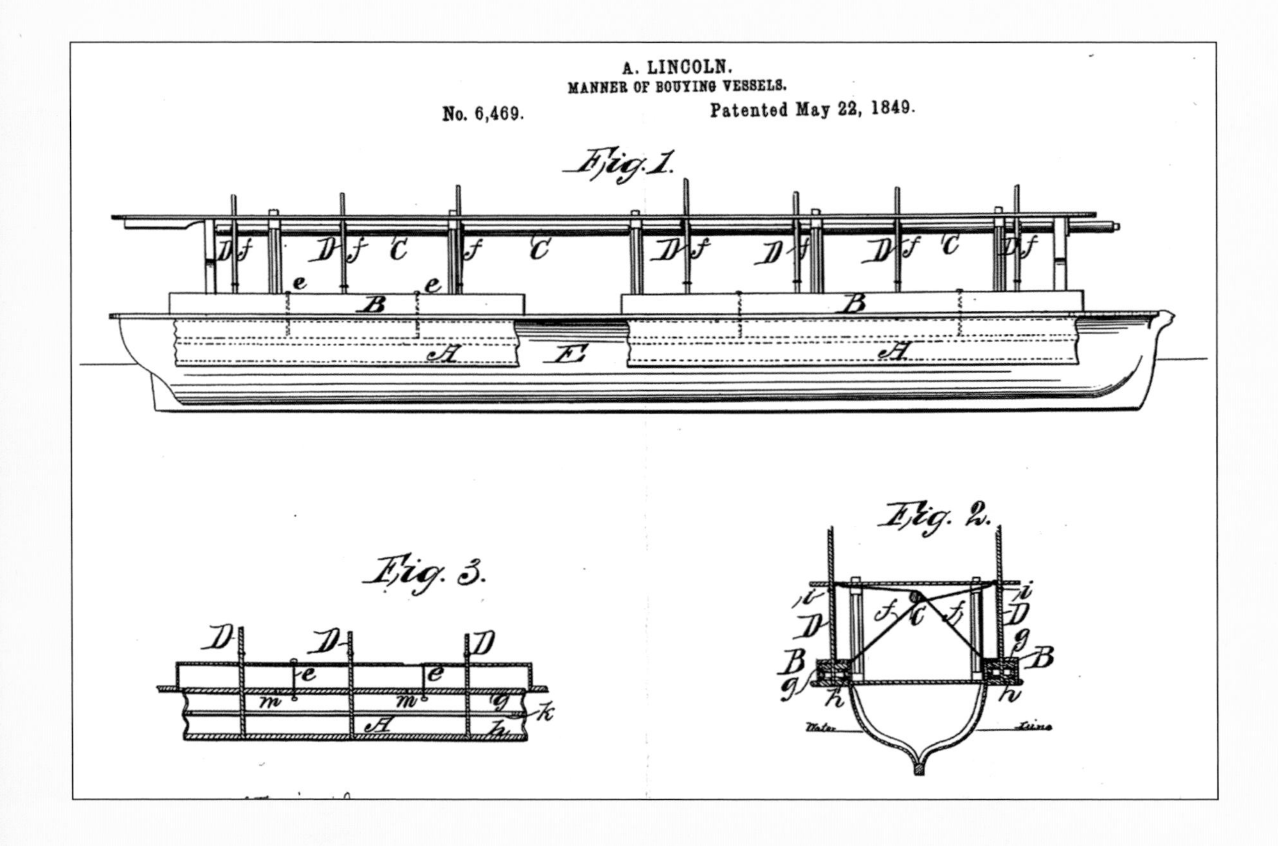

CHAPTER 10

GAS EXPLOSIONS: Safety Today a Product of Painful Past

The phoenix must
burn to emerge.
– *Janet Fitch, American novelist,* White Oleander

There is perhaps no scene more unsettling than a natural gas explosion.

Shredded personal effects, mortar, shingles, wood splinters, and the pungency of smoldering embers make for sobering imagery. Multistory residences in situ for decades are instantly dispatched, leaving but a crater as a kind of bookmark of the building's former address.

Nobody is certain as to what occurred March 2, 2011, at 885 Martin Road in Suffield Township, a community about 35 miles south of Cleveland.

It is assumed 63-year-old Regina Proudfoot and her 21-year-old grandson Robert Croft retired as they had often done on Wednesday evenings. But sometime between when they turned in and around four A.M., natural gas from a large outdoor propane tank had seeped into the living area.

It could have been either Ms. Proudfoot or her grandson who got up to use the bathroom. The simple act of flicking a light switch would have

House Ascending, *Painting by Ben Grasso.*

provided enough spark to ignite the gas. Or it could have been the gas found its way to the pilot light in the furnace. Cause of the ignition notwithstanding, the ensuing explosion would forever alter life on Martin Road.

The blast was heard by Suffield Fire Chief Bob Rasnick who was jostled from a sound sleep four miles away. Residents as far as eight miles from the explosion reported being woken from the discharge. One person 15 miles away said he could feel the rumble.

Back at 885 Martin Road, firefighters from six communities discovered a "ball of fire" fueled by remaining gas in the tank. As far away as 1,000 yards, Ms. Proudfoot's personal papers were found indiscriminately strewn. Fiberglass insulation, distinctively pink, was discovered in trees as were pieces of floor and roofing. The basement's structural beam was found almost 20 yards from where the house once existed.

Portage County Sheriff David Doak said authorities discovered the bodies of Proudfoot and Croft on opposite sides of the home's yard. It was widely believed death came instantaneously.

According to the Ohio fire marshal, there were 15 propane explosions since 2001 in the state resulting in five deaths. Almost a month earlier on February 7 in Hinckley Township, a community only 35 miles from Suffield, an elderly couple was also killed in their home by a violent propane explosion.

The National Fire Safety Association reveals nearly six million homes in the United States use propane gas. Nationwide, propane leaks caused 1,170 fires between 2003 and 2007, resulting in 34 deaths. What is not revealed in these statistics are the number of injuries and the resulting destruction of property.

Natural gas has been around as a fuel for almost 200 years. It is believed natural gas was first employed for illumination in Fredonia, New York, around 1821. It was later extracted from Pennsylvania oil and first used industrially in Pittsburgh before finding utility in other manufacturing centers.

While natural gas has significantly improved the well-being and lifestyle of a number of generations, it has been responsible for some extraordinarily gruesome catastrophes, the most devastating of which occurred March 18, 1937, at a high school in the East Texas oil field community of New London.

The time was several minutes past three and in just a few moments, fifth to eleventh graders would soon be making their afternoon exodus.

This school, by all accounts, was an attractive building. Framed of steel, the E-shaped structure had modernistic flair and was a source of pride for the residents of New London. And why not? Located in northwest Rusk County atop some of the richest oilfields in Texas, New London was among the most affluent rural school districts in the country.

But no one objected when school administrators—for purposes of controlling costs—decided to forgo a central steam heating plant or boiler room when the school was constructed in 1932. After all, the oil fields were flush with natural fuel and tapping a seemingly endless supply of residue gas would result in savings back then of $250 to $300 a month (about $4,100 to $4,900 today).

Use of the raw or waste gas was common in the oilfield community. Many among the populace took advantage of the free fuel, which was often burned in homes, churches and yes, even schools.

The method of distributing gas within the school was not unusual for the time and had been used at other educational institutions. It involved equipping each of the 72 classrooms with individual heaters.

Writing in the April 30, 1937, edition of the *National Fire Protection Association (NFPA) Quarterly,* H. Oram Smith of the Texas Inspection Bureau described the heaters thusly:

> The device has the appearance of an ordinary steam radiator, but is an individual heating unit comprising a gas burner at the base, under a small water chamber cast into the unit. Steam circulates through the hollow sections of the radiator and heats by radiation like the standard steam type.

Each unit had a small regulator at the source of the gas supply as well as a safety valve on the water chamber. Smith added: "It is a well-known make used extensively in the Southwest and is considered as safe as any gas heater on the market."

The residue fuel was delivered (via one-and-one-half inch pipes) to each of the rooms by a gas regulator connected to a two-inch pipe located in the school basement.

Fire officials speculated a large quantity of colorless, odorless gas collected in the basement area under an eight-inch slab of reinforced concrete serving as the main structure first floor of the two-story school. It was further theorized that at 3:17 P.M., a spark was generated when instructor of manual training Lemmie Butler plugged an electric sander into a receptacle on the first floor. Not unlike Martin Street, the worst school disaster in American history would forever alter life in New London, Texas.

Ignition of the gas produced a lone explosion with enough force to lift the school—including auditorium—off its foundation. The concrete slab floor was instantly catapulted through the roof by way of occupied classrooms.

While the detonation itself would have produced numerous casualties, falling debris from the slab and other building material further threatened building occupants. Many of the victims were crushed beneath masses of concrete, tile, and steel. Some surviving victims had to be extracted by jackhammer.

AY DROP TAKES 7TH AT TROPICAL

E WEATHER
g cloudiness tonight fol-
howers tomorrow. Con-
temperature. Moderate
esterly wind becoming
nd increasing.
ather Report on Page 44

TEMPERATURES.
Chicago 42
Omaha 44
Los Angeles 55
Salt Lake City........ 45
Seattle 43
New Orleans........ 67

THE BALTIMORE NEWS-POST
AN INDEPENDENT NEWSPAPER
The Largest Daily Circulation in the Entire South

XXX.—NO. 134. Entered as second-class matter at Baltimore Postoffice. FRIDAY EVENING, MARCH 19, 1937 PRICE 2 CENTS

Baltimore Widow Wins $150,000

BLAST KILLS 425

SWEEP WINNER AND GRAND NATIONAL FINISH

LLIAM J. SULLIVAN MRS. ELIZABETH MYERS MRS. MARIE MEEHAN

winner at Aintree! Royal Mail! And a fortune ped into the lap of Mrs. Elizabeth Myers, widow, h Fulton avenue, who held a sweepstakes ticket ector. Mrs. Myers is cheering at her good luck in the arms of two of her friends and neighbors. The ticket is worth $150,000, but it is reported that Mrs. Myers' attorney has sold a share of her ticket. Picture copyright, 1937, by The Baltimore News-Post. All rights reserved.

ROYAL MAIL WINNING GRAND NATIONAL

EEN (SECOND) DRIM (RIDERLESS) ROYAL MAIL

Workers End Search For Bodies In Ruins

NEW LONDON, Texas, March 19—(A. P.).—With 425 crushed and mangled bodies of children and teachers removed from the blast-shattered wreckage of the London Consolidated School, workers ended their search today for additional victims of the nation's worst school disaster.

Col. C. E. Parker, National Guard Commander, expressed belief every body had been found. A military inquiry to seek the cause of yesterday's great explosion quickly got under way within the ruins.

Every brick had been turned in the basement of the demolished building before workers pronounced their task finished, Colonel Parker said.

He declared about 425 bodies had been carried from the tangled pile of steel and brick and added:

"There may have been a few more than that."

Wearied oil field laborers, who had toiled with acetylene torches, hoisting machinery and bare hands for 21 hours, this morning in a slashing rainstorm, stopped work in groups and went to their homes, nearly all of which were bereft of at least one child.

The final hours of the tragic task showed a dismal drizzling scene, dotted with frenzied, red-eyed parents and determined investigators.

Major Gaston Howard, appalled at the nation's worst modern child tragedy, said an investigating board of six would start functioning this afternoon. Survivors and eye witnesses would be questioned in an open hearing, he said.

ACCUMULATED GAS BLAMED

First definite indication that accumulated gas caused the blast that lifted hundreds of school children, heavy girders and bricks into the air, came from Major Howard when he said Dr. E. P. Shoch, noted chemistry professor at the University of Texas, had been summoned to testify.

Rain slowed workers who had reached the basement of

Turn Over.

$150,000 WON BY WIDOW

Mrs. Myers, of Fulton Ave. Gets Fortune in Sweepstake Victory

Royal Mail's victory in the Grand National at Aintree tossed a fortune into the lap of an elderly Baltimore widow, Mrs. Elizabeth M. Myers, 1709 North Fulton avenue—but the amount of the fortune could not be ascertained immediately.

Mrs. Myers' ticket on Royal Mail won $150,000 but as her neighbors were celebrating her good luck in her home, the widow learned that a share of the tickt had been sold by her attorney.

OVER $100,000

At that, however, Mrs. Myers will receive over $100,000, her attorney said.

She will devote virtually all of her share, she said, to works of charity. And she would like to go to Ireland in a dirigible to collect the money, but probably will make the trip more prosaically aboard a ship.

Mrs. Myers' attorney, Judge Lee I. Hecht of the Appeal Tax Court, owns a share in her ticket, given him as a contingent counsel fee. Two other men, who Judge Hecht refused to identify, and who purchased a portion of the ticket own the other share.

Judge Hecht refused to say exactly what his share of the ticket

Continued on Page 36, Column 7.

PAVE WAY TO WALLIS DIVORCE

Court Rules Out Plea To Halt Decree; May Marry After April 27

By International News Service.

LONDON, March 19—(I. N. S.).—The royal road to romance for Mrs. Wallis Warfield Simpson and the Duke of Windsor was cleared of all obstacles today when the London divorce court threw out an application to block the Baltimore beauty's divorce on grounds of collusion.

Sir Boyd Merriman, president of the Divorce Court, announced his decision after Francis Stevenson, a Londoner who filed an intervention in the case last December, made known his willingness to drop the proceeding.

The action means that Mrs. Simpson and former King Edward VIII. of England may marry at any date after April 27, when the divorce decree nisi she obtained last October at Ipswich assizes becomes final and absolute.

Today's hearing in the ancient law courts where the strand runs into Fleet street, London's newspaper row, was held on a technicality.

Stevenson, an elderly, be-spec-

Continued on Page 8, Column 1.

Temperatures

2 A. M....	44	10 A. M....	52
3 A. M....	43	11 A. M....	53
4 A. M....	41	Noon	52
5 A. M....	40	1 P. M....	53
6 A. M....	42	2 P. M....	54
7 A. M....	43	3 P. M....	55
8 A. M....	46	4 P. M....	57
9 A. M....	49	5 P. M....	58

Fear Gas Peril In Oklahoma

OKLAHOMA CITY, March 19—(A. P.).—Worried parents beseiged school officials here tonight with requests to do something about the menace of gas pockets in the derrick studded East Side residential oil field district.

While a veteran fire fighter declared seeping gas was an ever-present oil field menace against which there was no adequate protection, school officials took extra precautions against the possibility of such a disaster as that which hundreds of school children apparently was caused by a seeping gas, some authorities said. There are seven oil wells on the school grounds.

C. K. Reiff, superintendent of Oklahoma city schools, ordered all school engineers to make a daily inspection for seeping gas here, where hundreds of oil wells fringe the city.

City Manager F. G. Baker was authorized by the Oklahoma City Council to increase his staff of oil field inspectors.

Thirteen Oklahoma city school

Very Latest News

(Race Results From Inter-State News Company, Inc.)

BAY DROP WINS SEVENTH AT TROPICAL

Seventh—Bay Drop, $51.10, $23.70, $12.50; Knights' Fancy, $9.70, $5.90; Busby, $5.90.

AT FAIR GROUNDS

Sixth—Gay Dog, $11.60, $5.20, $4.00; Persuader, $7.60 $4.20; Eniz, $3.40.

AT OAKLAWN PARK

Fourth—Bright and Early, $13.60, $6.10, $2.80; Transmutable, $4.00, $2.50; Appealing, $2.30.

AT EPSOM DOWNS

Fifth—Brown Prodigy, $4.80, $2.70, $2.50; Gallant Eagle, $2.80, $2.60; Miss De Mie, $2.90.
Sixth—Takus, won; Catchall, 2d; Dokas, 3d.

ESTIMATE BLAST DEATHS AT 500 TO 600

AUSTIN, Texas, March 19—(A. P.).—Capt. Walter Eliott of the State Highway Patrol reported from New London late today that "no two authorities agree as to the number of dead or the number unaccounted for," but "nearest estimate on deaths at present is between 500 and 600."

The Baltimore News-Post Today Is Printed In
THREE SECTIONS

Race Results At Tropical Park

CE—Purse $800; claiming; for three-year-olds; six fur- ff at 2.01. Time, 1.11 3-5.

ighland, 110 (Swain)	$15.40	$5.20	$3.80
ester, 116 (Westrope)		2.90	2.60
on, 112 (Le Blanc)			3.40

SECOND—Purse $800; claiming; for four-year-olds and up; six furlongs. Off at 2.31. Time, 1.11 3-5.

Kawagoe, 116 (Wright)	$33.10	$15.20	$7.40
Regula Baddun, 116 (Rosen)		5.30	4.20

Final death toll: 294 (and some believe as many as 319). This included 120 boys, 156 girls, four male teachers, 12 female teachers, a woman visitor and a four-year-old boy visitor. It should be noted no record exists of the actual number of people in the building during the incident. A total of 31 victims were sitting in the shop class when the gas ignited. Approximately 130 students were spared serious injury.

Hearing the blast, oil field roughnecks ran to the school site and toiled relentlessly to reclaim bodies and remove New London School's fragmented remains. Later that day, there would be over 1,500 volunteers on site from 20 organizations, agencies, and companies.

Firefighters arriving at the scene found no fire (there was a minimal amount of combustible matter) and so began the forbidding task of locating survivors and sifting through human carnage. According to the London Museum:

> Bodies were carried to hospitals in five counties. When those hospitals were full, they began to put bodies dead or alive in garages, American Legion halls, tents, churches, car dealer shops or any place that could be found...Word was spread for all doctors, nurses and embalming personnel.

At perhaps no time in history did more parents dread the horrible responsibility of locating deceased children, many of whom were mutilated and dismembered. The museum explained:

> Horror-stricken and agonized families rushed to the scene frantically searching for their children through the mounds of rubble with tears running down their faces and hands torn and bleeding from jagged debris.

This was not the life cycle as intended: Children preceding parents in death. The few who survived were located at remote ends of the building structure opposite the explosion site.

Damage was not limited to the immediate school area. A 1936 Chevrolet 200 feet from the scene was crushed by a two-ton slab of concrete. Another 50 vehicles were totaled after being struck by airborne concrete and stones. Added the museum, "Some of the flying wreckage included precious children thrown through the air like broken rag dolls."

As with any disturbing event of this magnitude, news reporters from all over the state headed to New London. A cub reporter from the United Press International office in Dallas was excited about covering his first major story. As he arrived on the scene, he observed floodlights being readied for the evening darkness, as well as erection of large oil field cranes that would be used to assist the removal of large hunks of rubble.

Years later as this reporter approached the twilight of his career, he observed, "I did nothing in my studies nor in my life to prepare me for a story of the magnitude of that London tragedy, nor has any story since that awful day equaled it."

Despite the fact that he expertly chronicled every major story from all over the world during an iconic 45-year broadcasting career, Walter Cronkite could never dislodge the recollection of the horrible scene he personally witnessed on March 18, 1937.

Less than a day following the explosion, the New London School site was completely devoid of any and all remnants from the previous day. Reported the NFPA *Quarterly*:

> In the short space of 17 hours after the work was organized, some 2,000 tons of debris were picked up piecemeal and hauled away during an all-night rainstorm; concrete slabs were broken up, tangled steel cut with torches and the smaller fragments that had to be shoveled were carried off in small baskets and carefully emptied under flood lights to avoid overlooking a hand or foot or any torn portion of a body.

There was no shortage of theories on the cause of the school disaster. Unfortunately, in the accelerated confusion to save lives and remove rubble—not to mention the dreadful memories—there would be no evidence to dissect. Much of what would be learned later was based on conjecture, albeit conjecture having some foundation in logic.

Simply explained: "*This method of heating was entirely wrong and in combination with the unventilated floor space was responsible for the explosion,*" wrote Texas Inspection Bureau's Smith. Cause of the explosion, he noted, was ignition of a sizable gas pocket in the large improperly vented space under the floor.

Smith explained the space under the concrete floor became filled with an explosive mixture of gas and air. (Source of the gas leak was undetermined.) He concluded the gas found its way into the school shop area by way of an open door and was detonated by an arc formed when the sander plug was introduced into the receptacle. The consequent flash retreated under the concrete floor, creating superheated gas and enough pressure to dislodge the school from its foundation. The blast traveled the building's entire 254-foot length.

Following an investigation, a court of inquiry exonerated all school officials of the explosion and concluded no one individual was responsible.

Thirty New London high school seniors who survived the blast finished their academic year in temporary facilities as a new school was constructed on nearly the same site as its doomed predecessor.

As bad as this human catastrophe was, it could have been—although difficult to believe—much worse. The school had been using dynamite at its athletic field to build a running track. At the time of the blast, 18 sticks of the ordnance were stored in a lumber room under the auditorium. None was disturbed.

There is little solace in discovering something positive resulting from such a horrendous turn of events. But the deaths of 276 children—nearly a complete generation of the New London community—did result in something that has saved the lives of perhaps millions of people worldwide.

Up to this point in history, gas was clear, odorless matter. But within weeks of the New London incident, the Texas legislature passed an odorization law requiring the addition of distinctive malodorants to all gases used commercially and industrially. It was not only the first law of its kind, it is currently law throughout the United States.

Since the raw unprocessed gas at New London School was being tapped directly from its underground source, there probably was no mechanical procedure to odorize the gas mixture collecting under the concrete floor. Hence, it is doubtful the 1937 disaster could have been averted.

Today, however, a smell tantamount to "rotten eggs" serves to alert anyone in close proximity to leaking natural gas and the real potential for an explosion.

Despite new technology, gas explosions still occur with alarming regularity. And that's why codes and standards remain critical to the well-being of every man, woman, and child.

As H. Oram Smith concluded in his review of the New London tragedy:

> Practically all faults of construction and installation in this building were due to lack of supervising power such as would apply in communities having city ordinances. It serves to focus attention to the need of state laws on standards of construction, as well as approved standards for the installation of heating systems, electrical equipment, gas and oil systems and all other buildings where large numbers of people congregate.

And that's the way it is.

SCENES FROM THE NEW LONDON SCHOOL EXPLOSION

(Warning: graphic images)

All Photos Courtesy of London Museum.

Dazed unidentified woman stares blankly into the school ruins.

Volunteers remove the dead.

Volunteers rush to extricate victims.

Crowds gather at the explosion scene.

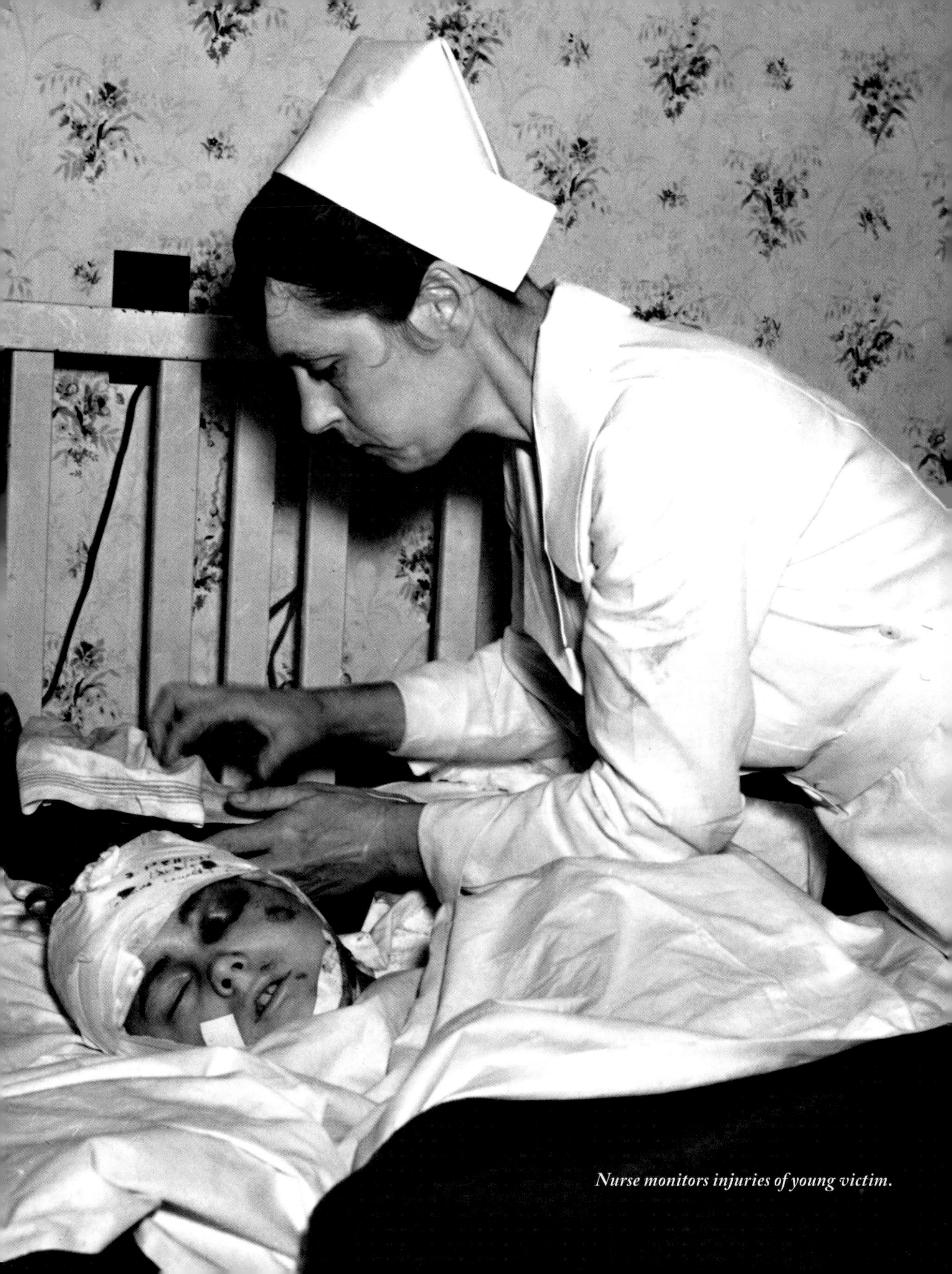

Nurse monitors injuries of young victim.

Aerial view of the destruction.

One of many vehicles crushed by flying debris.

Workers carefully pick through debris.

Young victim trapped among the rubble.

Volunteers work through the night to purge the school site of heartbreak.

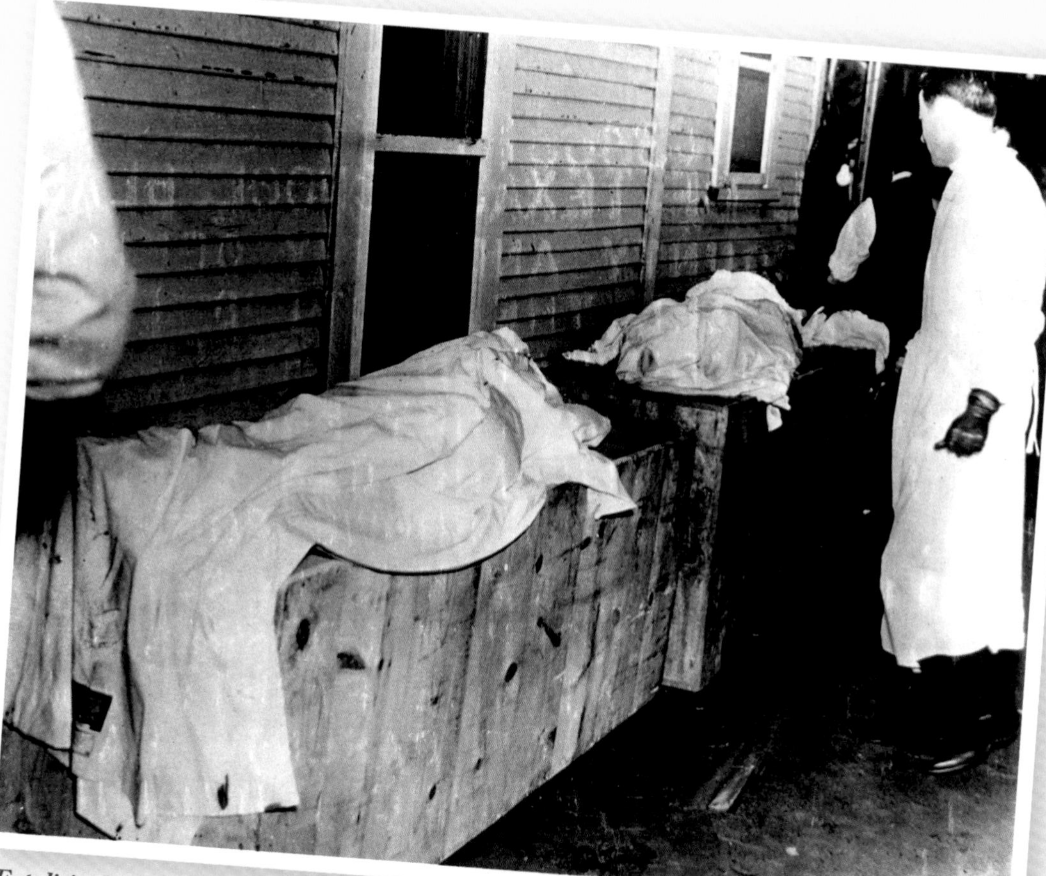

Fatalities lined up for identification.

Transporting a victim's casket.

New London School hearing: police official examines classroom heating unit.

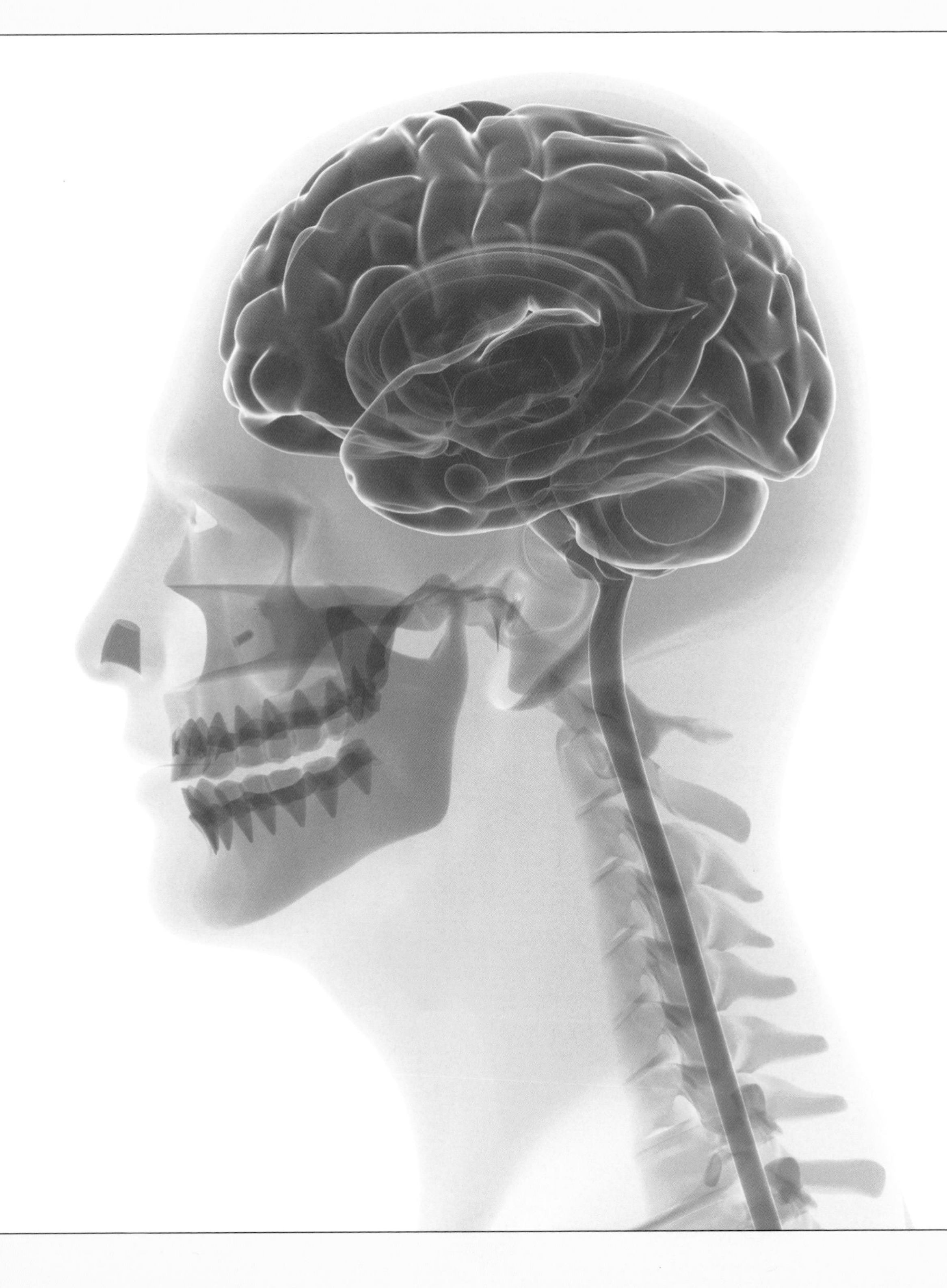

CHAPTER 11

PRESSURE EQUIPMENT ACCIDENTS: Communicating the Human Toll

There's nothing that cleanses your soul like getting the hell kicked out of you.
– WOODY HAYES, AMERICAN FOOTBALL COACH, 1913–1987

[This essay by the book's author was adapted from a column published in the Summer 2008 National Board BULLETIN*]*

IT HAS BEEN LONG HELD grisly photographs make a significant safety statement.

Such was the theory of several safety organizations during the 1950s and 60s looking to reinforce their badgering messages of caution and vigilance. Stark images of death, they somehow reasoned, were effective deterrents to reckless behavior.

And there were other outlets for those seeking to feel better about themselves while spying the intimate misfortune of other poor souls.

Pictures of car accident victims and those killed by catastrophic events were routinely published in early twentieth century newspapers. Not only did this feed public curiosity, it sold newspapers.

By today's standards, making these types of visuals available to the public is considered bad form. Fear has been replaced by education as the most effective way of reaching and, yes, persuading what is often a skeptical public, i.e., focusing on *prevention* of accidents, not the *consequences*.

However, despite the absence of graphic photos, there is no rationale *not* to educate those seeking a more detailed explanation of accident reality: that is, revealing—via the written word—the horrible human suffering that may occur when a person is within close proximity of either an equipment explosion or rupture.

So without getting too graphic, I have cobbled together—with the help of a couple of outside sources—a short primer on what happens to a body when a catastrophic release of pressure ensues.

CAUSE AND EFFECT

It can be correctly assumed concussion (i.e., shock wave caused by impact) resulting from an explosion is responsible for most deaths and injuries. But is flying debris always the primary cause?

In his November 2000 presentation *Defining Disaster: The Emergency Department Perspective*, noted physician Dr. Joseph Zibulewsky of the Department of Emergency Medicine at Baylor University Medical Center observed:

> Most blast fatalities are from brain injuries, skull fractures, diffuse lung contusions, and liver lacerations . . . Only about 15% of those who come to the emergency department for blast injuries are admitted to the hospital; others are either well enough to go home or do not survive the initial blast.

Dr. Zibulewsky explained the initial pressure wave of a high-energy explosion travels at 800 meters per second. The body parts most affected by a blast wave, he noted, are lungs, ears, and gastrointestinal tract. Injuries are caused by a series of "mechanisms."

The first involves the movement of particles in the body from more dense areas to less, as when liquid in the lungs is displaced and causes pulmonary hemorrhage. With pressure differentials, the blast wave forces fluid from its space, which can cause death hours or even days after the explosion. Compression and decompression can also rupture membranes in the ears.

"Among secondary injuries are cuts caused by flying glass, shrapnel, and debris that can imbed deeply into tissues," Dr. Zibulewsky expounded. "Tertiary injuries occur as people are thrown against hard surfaces."

Doctors say most surviving explosion patients fail to remember anything relating to the blast concussion. As a child, I was struck by a car and knocked unconscious. There was no pain. And to this day, I have no recollection what happened in the minutes preceding the accident.

Going into shock, doctors explain, is the brain's way of dealing with a traumatic event.

STEAM CAN BE DETRIMENTAL TO YOUR HEALTH

Although concussion from a pressure equipment blast is the foremost cause of death and injury, steam—or more precisely, superheated steam—is the most feared by those understanding its destructive potential.

One of the oft-told anecdotes in our industry is of utility operators using a broomstick to check for high-pressure steam leaks. Because superheated steam can sever appendages, it was believed passing a broomstick over a suspected leak area was preferable to any alternative involving truncation.

While this makes for a good story, I have been unable to find anyone in the utility industry who subscribes to the broomstick practice. The reality is *any* leak of superheated steam will produce a significant sound, not unlike an air horn, that can be distinctly heard throughout many generating plants.

According to columnist Cecil Adams, the broomstick practice may have started aboard ships with steam boilers many years ago. "Tight quarters and noisy conditions could have made it difficult to find leaks quickly without some direct evidence," he speculates.

Although the broomstick story may be only an old wives' tale, the dangers of superheated steam are not.

Searing heat notwithstanding, any blast of high-pressure gas can render significant physical damage. Adams cites US Army medical reports describing a number of high-pressure gas injuries. "Just 12 psi can likely pop your eyeball from its socket," he points out. (The U.S. Occupational Health and Safety Administration limits air pressure used in industrial cleaning to 30 psi.)

As we all know, a high-pressure leak is dangerous for a variety of reasons, including turning any loose objects into flying projectiles. Tools, tanks, lunchboxes—anything—being jettisoned at a high rate of speed can be lethal to the unsuspecting.

Those lucky enough to escape being struck may not be as fortunate when dodging a steam leak, particularly in confined spaces.

"Even at lower temperatures enough steam in a small area can suffocate you as it displaces air," Adams writes. "A big steam leak . . . can quickly raise the surrounding air temperatures so high you'll cook from the inside if you breathe."

There are perhaps few who can better describe the internal and external effects of steam than Dr. Michael Badin, respected pathologist whose expertise is frequently sought on TV news and reality programs.

"A steam burn does not singe the hair or char the skin," notes Dr. Baden. "Steam has a temperature of 212°F or more and can cause burns of the outside of the skin. In the steam deaths that I've seen, the causes of death have largely been burns of the air passages (the mouth, the windpipe, the air passages in the lungs). When superheated air, vapors and steam are inhaled into the body, the exposure can cause very rapid death."

The media often wrongly equate steam and smoke inhalation, the difference being the quantity of heat carried by steam. (Some studies indicate steam's heat-carrying capacity is 4,000 times greater than hot air.) Industry authorities say humans can withstand air temperatures and dry steam up to about 200°F. Conversely, hot condensate can generate pain (burning of the skin) at 120°F. As air is displaced and the amount of steam by volume increases above 12 percent (not unusual in confined spaces), temperatures exceed 120°F and, well . . . there is seldom a happy outcome.

Those fortunate enough to survive face years of medical intervention and disfiguration both externally and internally. But there is yet another injury victims might endure: post-traumatic stress disorder brought about by stress and shock.

Sound horrible? You bet.

The next time you read about a person killed or injured in a pressure equipment accident, reflect on the victim and what he or she endured. With disturbing images all around us on television and in movies, one doesn't require much imagination.

Or a photograph . . . ✲

CHAPTER 12

WHEN IS A BOILER EXPLOSION *NOT* A BOILER EXPLOSION?

He'd been wrong, there was a
light at the end of the tunnel,
and it was a flamethrower.
– *Terry Pratchett,*
English novelist, Mort

[This essay by the author was adapted from
a column published in the Summer 2007
National Board BULLETIN*]*

I HAVE A THING about words. One of my favorites is *verisimilitude.*

Quite simply, verisimilitude means the appearance of truth.

Example: If you believe everything you read, natural gas explosions are often caused by faulty boilers or water heaters. The fact it has been reported so many times lends a certain credibility to what is in reality a leap of significant expanse. That credibility is verisimilitude.

Gas *is* used to fuel pressure equipment. But does that mean a flawed boiler or water heater *causes* a gas explosion? Industry purists argue gas leaks

Photo by Ben Bailey.

are a consequence of using pressure equipment, but seldom the root cause of a boiler explosion.

So why are so many gas explosions reported as malfunctioning pressure equipment?

Problems — and explosions — occur when escaping gas collects in a confined space and is ignited by a furnace employed to heat the water in the boiler.

The difference between furnace and boiler explosions is unmistakable: A boiler explosion occurs when the contained water and/or steam is suddenly released to the atmosphere. The result is a lethal shockwave instantly dispatching shards of metal and scalding steam — *but with no resulting fire*. Furnace explosions occur when a furnace ignites a pocket of confined gas, thereby provoking fire and blowing everything — everything — to hell.

That said, how can it be argued explosions involving leaking gas are genuine *boiler* explosions? After all, insurance companies distinguish furnace and boiler explosions by writing fire policies for the former and machinery policies for the latter. Another rationale for differentiating the two: A furnace explosion can cause a boiler explosion. It is extremely unlikely a boiler explosion can cause a furnace explosion.

Given a lack of understanding (as well as its pursuit of simplicity and convenience), the media is wont to call any explosion within close proximity of a boiler, "a boiler explosion."

This point is perhaps best validated by the catastrophic "boiler explosion" at Michigan's Ford Rouge plant in 1999. That tragic incident was prompted by a buildup of gas in the boiler plant. Although the gas accumulated in the plant and not *inside* the boiler (which was shut down for maintenance), it was — according to the media — "the Ford Rouge *Boiler* Explosion."

That dreadful event, resulting in six fatalities and 20 injuries, was caused by "inadequate controls for the shutdown of the boiler," according to the state inspection report. Translation: There was nothing a boiler inspector could have done to pre-empt this tragedy of errors.

The Rouge incident lends perspective to yet another distorted perception that regularly confounds purists: the boiler room as a location of ominous repute. Indeed, being a boiler's address of record, it stands to reason boiler accidents occur in, well, the *boiler room.* But accidents do occur even without a boiler room.

One of the twentieth century's most devastating accidents occurred March 18, 1937, in the East Texas oil field community of New London. The New London School, as it had always done, tapped into a residue gas line through its basement to heat classrooms with individual boiler-type steam radiators (72 in total) — fueled by separate connectors. A leak in the basement allowed colorless, odorless gas to collect in the basement area, mix with air, and seep into the school wood shop located on the first floor. A spark generated when the shop instructor plugged an electric sander into a receptacle creating an explosion that literally lifted the high school — including auditorium — off its foundation.

Impact of the blast launched the main structure floor (an eight-inch concrete slab) through the roof by way of occupied classrooms. Many of those killed were crushed under falling debris. A total of 294 students (fifth to eleventh grade) and teachers perished.

Some published accounts to this day refer to an explosion in the boiler room. But according to a 1937 National Fire Protection Association report, New London School had no central steam heating plant or boiler room (both casualties of efforts to control construction costs).

So oft repeated was the boiler room reference that it today remains gospel for many. Hence, another example of the evil-boding boiler room *and* the boilers occupying them.

These are but two examples of verisimilitude. Others are too numerous to recount. And that is a problem.

Many would reason that whether a furnace or boiler explosion, such accidents underscore the dangers of pressure equipment. So why should we care?

According to the World Health Organization, the number of boiler explosion/rupture fatalities in 2004 worldwide was 51. The United States reported only three deaths with Mexico (five) and Brazil (eight) recording the most. While the United States did not have the lowest tally of fatalities, it did boast — more significantly — the lowest number of deaths per capita (.0101443 per million) among 26 industrialized nations.

Truth is, the pressure equipment industry should be proud of its safety record. However, as one industry pundit reminds me: We are only judged by our failures. But an overwhelming percentage of those failures had nothing to do with the inspection process. Most accidents are the result of human error — committed by humans having no connection to inspecting boilers.

Nonetheless, the boiler industry is generally guilty by association with all things bad in the boiler room. Reality be damned.

The industry recognized this a long time ago. Fearing a loss of something in the translation, it advanced an aggressive agenda of *preventing* accidents rather than *promoting* accident consequences.

Regrettably, the Rouge Plant incident will be forever classified a boiler explosion, at least by those not in the know. But, fortunately, there are those who accurately chronicled the real cause of the New London disaster. And for good reason.

Following this incident, the state of Texas enacted what is believed to be North America's first odorization law requiring natural gas to be mixed with distinctive malodorants to detect gas leaks by smell. Even the Rouge disaster caused Ford officials to redouble safety efforts in such a way that Ford plants are today considered to be among the safest in the automotive industry.

The next time you hear about a boiler accident, think verisimilitude. And then identify the cause. The real cause.

While good occasionally arises from the aftermath of an explosion, the end result for the pressure equipment industry is rarely positive.

A *failure* is a *failure*, notwithstanding what the public chooses to call it.

CHAPTER 13

POSTCARDS FROM HELL

The darkest hour has
only sixty minutes.
– Morris Mandel, American-
Jewish educator,
1911–2009

Through the years *we have all received colorful postcards displaying a beach scene, a famous building, or maybe even a spectacular sunset. But how many can truthfully admit to having been a recipient of a black-and-white postcard featuring—of all things—the mangled remnants of a boiler explosion?*

While appearing a bit distasteful, such communiqués were quite common in another time and place.

In the late 1800s and early 1900s, boiler accidents, like many disasters, were not only spectacular in their aftermath, they were oftentimes newsworthy. Sending a postcard to a friend or loved one featuring such a forbidding incident was often a way for residents of the town in which the event occurred to share a bit of community notoriety, albeit in a rather ghoulish way. After all, many of these accidents involved the deaths of local residents: sometimes friends, most certainly neighbors.

What follows is an absorbing mosaic of past images carefully culled and assembled as a kind of media primer on boiler and pressure vessel safety.

Today, these photographic pasteboards serve not only as a reminder of the way things were, but perhaps more important, as a sort of historical documentation of why pressure equipment needs to be inspected.

In retrospect, examining these historical picture cards reflects a certain finality. In the truest sense, they are a glimpse of a moment captured in time—an unvarnished reflection how disaster was handled by another generation. The expressions of the people in the photos are often indifferent, almost disassociated from the reality that surrounds them. And then there are the public officials who, by dint of their jurisdictional authority, strike a formal pose for the record, and subsequently, posterity.

Unidentified and undated boiler explosion believed to have taken place in Québec, Canada.

Unidentified and undated boiler explosion in Cuba, New Mexico.

R-M CARD CO. -3
SHOEPKE HOME - DAMAGED BY THE
KEMPF BOILER EXPLOSION - MAR. 17 '08. - CRANDON, WIS.

KEMPF
PLANING-MILL
3-KILLED
-INJURED
MAR. 17-1908-
8 A.M.
A.M. CARD CO.
- THE RESULT OF A BOILER EXPLOSION -
CRANDON-WIS.

-BOILER-
EXPLOSION
AT THE KEMPF
PLANING MILL
MAR, 17-08
8-A.M.
BOILER WAS
BLOWN NEARLY
TWO BLOCKS FROM MILL-
CRANDON-WIS.
A-M CARD CO.
PHOTO-

Boiler accident at the U.S. Gypsum Company, October 16, 1910.

RUINS OF BOILER EXPLOSION IN THE LAKEPORT
STEAM LAUNDRY. Laconia, N. H. July 5, 1910.

Boiler Explosion, Westville, N.S., April 2, 1914.

Wreck of Electric Light & Water Plant. Robinson, Ill.

YORK ROLLING MILL, AFTER THE BOILER EXPLOSION. 10 KILLED, 19 INJURED.

Boiler explosion, So. Boardman, Michigan, April 4, 1912.

CHAPTER 14

DESTINATION: Destruction

The ultimate result of shielding men from the effects of folly is to fill the world with fools.
– Herbert Spencer, English philosopher, biologist, sociologist, 1820–1903

As hard as one looks, it cannot be found on any map.

Crush, Texas, existed for only a day. But its claim to notoriety involved one of the most unusual, if not deadly, publicity stunts ever executed. And in terms of railroad lore, the Crash at Crush transcends legend.

America was in the throes of economic depression in 1896, with businesses of all types struggling to exist. Railroads were not immune.

One such railroad was the Missouri Kansas & Texas line or as it was often called, the M.K.T. or Katy Railroad.

To counter the reduction in rail traffic, the Katy Railroad hired William G. (Willie) Crush as assistant to the line's vice president. And for good reason. Crush brought to the railroad a set of promotional skills engendered by his association with one of the all-time great showmen, P.T. Barnum.

Willie Crush envisioned his job as drawing attention to the Katy line, with a hope such exposure would enhance rail traffic.

What better way, he thought, than a train crash. A *staged* train crash. After all, people come from miles around to witness tragic events. (A staged crash several months earlier by the Columbus and Hocking Valley Railroad near Cleveland had attracted 40,000 onlookers. However, it is unclear whether Mr. Crush was aware of the event.)

Crush's proposed crash was rather simplistic: Choose an out-of-the-way location, promote the head-on wreck of two old train engines, charge spectators $2 train fare to visit the event, and conduct all of it within a festive, carnival-like atmosphere.

Willie's excitement to orchestrate such a truly bizarre promotion was mirrored by his Katy Line superiors.

Naturally, the first order of business was to designate a location. A desolate area 15 miles north of Waco seemed to be an ideal setting for most Texas residents to access. In preparation, 500 railroad workmen laid a four-mile spur alongside Katy Railroad tracks.

As locations for train wrecks go, the "City for a Day" called Crush could not have chosen a more natural spot. Enclosed by hills rising on three sides, the special track was situated in a shallow valley, thus creating an almost perfect amphitheater.

Also constructed at the site were a grandstand, three speaker stands, a bandstand, two telegraph offices, and a stand for reporters. Two wells were sunk as water sources for the large number of faucets needed to quench the public thirst. To formalize the location, a depot with a 2,100-foot platform was constructed along with a massive sign welcoming visitors to Crush, Texas.

Securing the two trains to be sacrificed was of little consequence. The railroad's upgrade to 60-ton steam engines from 30 tons made the latter obsolete and easily expendable.

Three of the engines were selected for the event: two for the controlled collision and another as backup. One of the rail vehicles was painted red with green trim while the other was decorated green with red trim. Mechanics devoted countless hours to preparing the two engines for upcoming publicity events.

During the scorching summer of 1896, the Katy Railroad commenced distribution of posters and circulars announcing the head-on "Monster Crash" of two steam engines traveling full throttle. Every telegraph pole along the Katy line from Missouri to Kansas to Oklahoma and Texas featured promotional handbills. Publicity for the upcoming crash reached far across the Texas plains as some newspapers chose to provide daily articles on event preparations. Newspaper ads aplenty greeted daily readers.

As for engine "999" and engine "1001" (so adorned by the promoters), the publicity tour was an outstanding success all along the Katy line, with thousands of the curious and concerned seeking to steal a glance at the two aging behemoth engines.

And then came September 15, 1896. On the day of the event, a circus tent would serve as a restaurant, while a large carnival midway complete with dozens of game booths, soft drink stands, and medicine shows would entertain legions of excited spectators.

Willie Crush's abilities to attract the curious were nothing short of genius. Folks all along the Katy line made their way into passenger cars so full that many were forced to ride on top of the cars. By mid-morning, it was estimated a crowd of more than 10,000 had gathered. At noon, the enthusiastic audience tripled. And shortly before the event, attendance was estimated to have climbed to between 40,000 and 50,000.

With a crowd tantamount to the population of a large Texas city at that time in history, Crush was well prepared. In addition to structures built earlier, a wooden jail was constructed to detain those having questionable agenda or character. Of course, that required some form of law enforcement. Between 200 and 300 constables were dispatched to the temporary community and proceeded directly to the task of escorting drunks, pickpockets, and troublemakers to the new detention quarters.

While spectators amused themselves along the makeshift midway, crew engineers worked on the featured performers: two 35-ton Pittsburgh 4-4-0s each pulling six cars displaying advertisements for Ringling Bros. Circus and Dallas' Oriental Hotel. Tests were conducted to estimate the precise location of impact.

Not all those preparing to witness this remarkable event were enthusiastic about its outcome. Railroad officials expressed concern for the boilers on each train. It was noted the impact could rupture one if not both and thus unleash an incredible quantity of steam (1,600 times the volume of water contained therein). Steam pressure, after all, was such it could move gigantic pistons that in turn provided locomotion of the engine and trailing railroad cars.

To placate his critics, Willie Crush conferred with train engineers, most of whom assured him neither boiler would rupture as a result of the violent impact. A minority opinion was rendered by an old engineer named Hanrahan who worked on railroad lines in America and Ireland. The veteran railroader unapologetically offered: "They'll [the boilers] burst and kill people all over the place."

Sensing the rewards and satisfaction of a successful albeit unusual event, Willie Crush was not deterred. Hanrahan's concern was disregarded, and a favorable report regarding the safety of the engine boilers was passed along to Katy officials.

Satisfied with Willie's explanation, railroad executives agreed to proceed with the event.

The scheduled four P.M. collision was delayed an hour to permit scores of additional railroad passengers to arrive.

The crowd took position on a hill 200 yards from the collision point. The September 14 *Dallas Morning News* described the location as having "a perfect view of the destruction." Photographers and reporters were allowed within 100 yards of the event.

At five P.M., the two multicolored engines with rail cars in tow ceremoniously met cow-catcher to cow-catcher at the soon-to-be collision point where photos were snapped commemorating the occasion. Subsequently, the trains slowly backed away from each other to starting locations two miles apart.

At approximately ten minutes after five, Willie Crush mounted a white horse, alerted telegraphers to signal the trains, and raised his hat to the enthusiastic roar of spectators. Engines "999" and "1001" embarked on their last journeys.

They would last only two minutes. And after achieving a speed of 45 miles per hour, the crews of both trains tied the throttles open and jumped from their engine positions. The September 16 edition of the *Morning News* described what happened next:

> The rumble of the trains, faint and far off at first, but growing nearer and more distinct with each fleeting second, was like the gathering force of a cyclone. Nearer and nearer they came, the whistles of each blowing repeatedly and the torpedoes which had been placed on the track exploding in almost a continuous

The "handshake." Both trains met by touching cowcatchers at what would be the point of collision before each backed one mile to its respective starting point. Both engines hauled six cars, two of which on each train carried advertisements for Dallas' Oriental Hotel and, appropriately enough, Ringling Brothers Circus. COURTESY: The Texas Collection, Baylor University, Waco, TX.

> round like the rattle of musketry . . . . They rolled down at a frightful rate of speed to within a quarter mile of each other. Nearer and nearer as they approached the fatal meeting place the rumbling increased, the roaring grew louder . . . .
>
> Now they were within ten feet of each other, the bright red and green paint on the engines and the gaudy advertisements on the cars showing clear and distinct in the glaring sun.
>
> A crash, a sound of timbers rent and torn, and then a shower of splinters.
>
> There was just a swift instance of silence, and then as if controlled by a single impulse both boilers exploded simultaneously and the air was filled with flying missiles of iron and steel varying in size from a postage stamp to half of a driving wheel.

During a surreal moment, disbelief collided with reality.

While the engines smashed together as planned, rupture of the boilers was not anticipated.

The violent, indiscriminate launch of wood and metal projectiles provoked instant panic. Spectators ran as though their lives depended on it. And indeed they did.

Standing shoulder to shoulder, most of the crowd was unable to move, let alone dodge the shower of scalding shrapnel.

The trains just before moment of impact. COURTESY: The Texas Collection, Baylor University, Waco, TX.

Ernest Darnall, who watched the event sitting in a Mesquite tree, was killed instantly when a hook attached to a wrecking chain struck him in the forehead and cracked open his skull.

Jervis C. Deane, the official event photographer, was struck by an airborne bolt that ripped out his right eye. The pressure wave also left the Waco resident's head peppered with small metal shards. Photographer J. Louis Bergstrom was struck unconscious by a wood plank.

DeWitt Barnes of Hewitt, Texas, was struck and killed by a soaring equipment fragment while standing between two women, one of whom was his wife. Neither of his companions was injured.

Countless others were either scalded or struck by hot debris. A Civil War veteran likened the scene to a battlefield. Two trucks, each weighing a ton, were lifted from the ground and tossed end over end a distance of 300 yards.

Three people died. Scores of others were injured, six seriously.

When the shower of shrapnel concluded, spectators ran toward the wrecked colossal machines to collect souvenirs. A number of people were burned as they attempted to retrieve equipment parts that remained hot to the touch.

According to the *Morning News* :

All that remained of the two engines and twelve cars was a smoking mass of fractured metal and kindling wood, except one car on the rear of each train, which had been left untouched. The engines had both been completely telescoped, and contrary to experience in such cases, instead of rising in the air from the force of the blow, were just flattened out.

Katy officials wisely ascertained a number of lawsuits would be filed forthwith. Before sundown and as quickly as his festive event turned tragic, Willie Crush was fired.

Cause of the exploding engine boilers was never officially determined. But it is more than

Like children at a circus, excited onlookers rush to the site of impact to secure crash souvenirs still hot to the touch.
COURTESY: The Texas Collection, Baylor University, Waco, TX.

The dramatic collision. Ragtime composer Scott Joplin would later pen "The Great Crush Collision March" to commemorate the deadly stunt. COURTESY: The Texas Collection, Baylor University, Waco, TX.

conjecture to speculate the collision of two train engines traveling toward one another at a speed of 45 miles per hour may have been a factor.

The Katy line responded to those injured and families of the deceased by quickly providing financial settlement. Photographer Deane received $10,000 (about $270,000 in 2012 dollars) and a free lifetime pass on the Katy line as compensation. "Having gotten all the loose screws and other hardware out of my head," he told Waco newspapers, "[I] am now ready for all photographic business."

The storied Crush, Texas, was located near McLennen County, three miles west of Interstate 35. Aside from a roadside marker, not even a hint of the event that took place 115 years ago remains. The "City for a Day" is now cow pasture.

And Willie Crush?

The unintended consequences of September 15 yielded their intended results. Passenger traffic on the Katy improved significantly as news of the debacle circulated around the world. Within a day of his firing, Willie Crush was discreetly rehired.

After 57 years with the Katy line, he retired in the early 1950s.

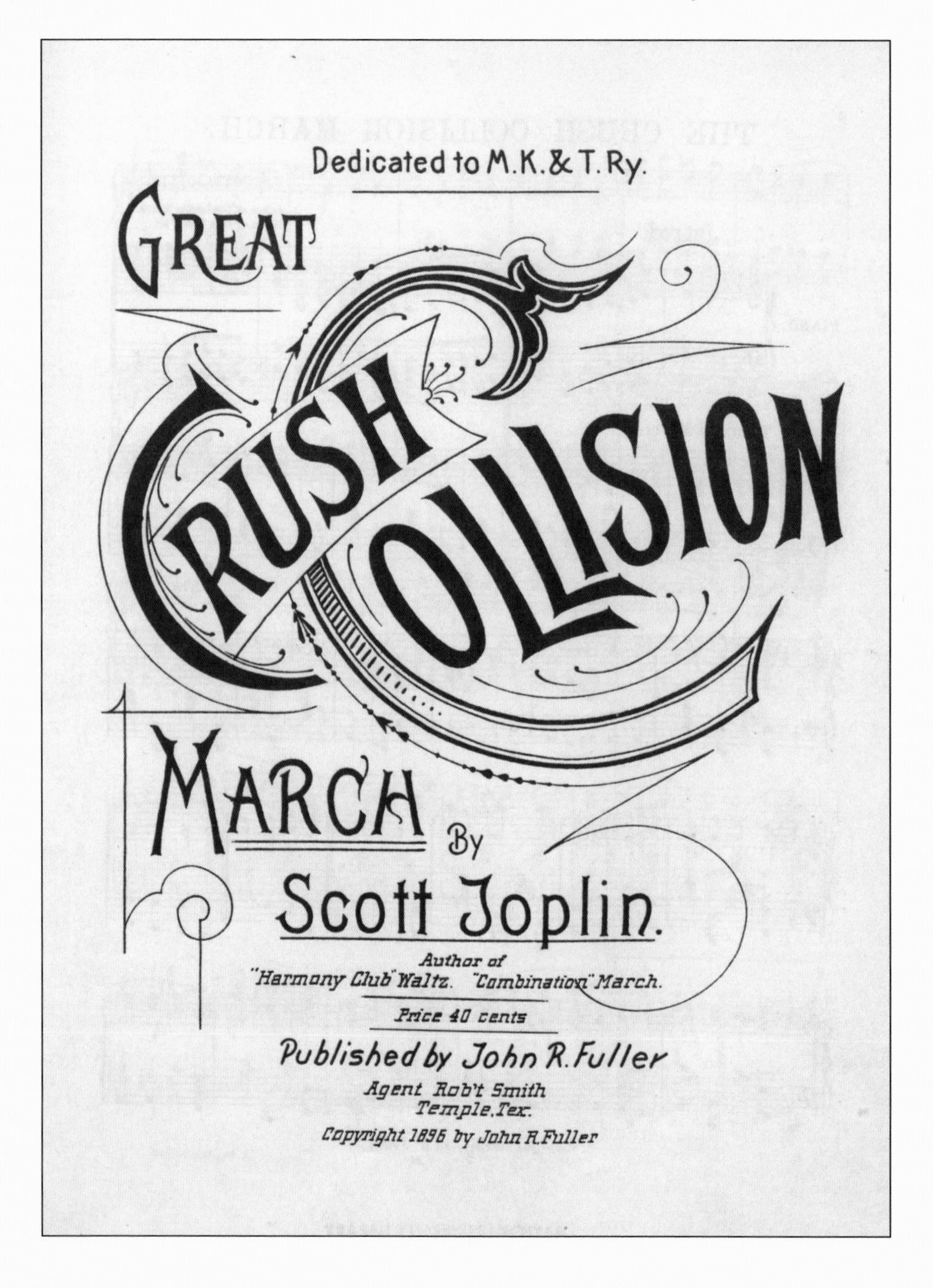

CHAPTER 15

STEAM LOCOMOTIVES: A Primer in Pictures

If you're going through hell, keep going.
– *Winston Churchill, English statesman, 1874–1965*

For over a century, *they were symbols of America's industrial might.*

Traversing the country like one-eyed behemoths, steam locomotives not only moved a country's hopes and dreams, they transported the people who would make those hopes and dreams reality.

And they were powered by boilers. Massive boilers.

What steam locomotives so impeccably illustrated—in both symbol and substance—were the countless contributions of boilers to the evolution of a great industrialized nation.

Indeed, without steam locomotives, America's growth could have easily evaporated. How else could a vast, burgeoning country have relied upon limited proximity to navigable waterways or seacoasts? How could important industrial and agricultural markets expand to meet the needs of a newly expanding populace? How could massive quantities of goods and materials be moved economically and quickly? How could people—and yes, news and ideas—be transported conveniently and efficiently?

For recent generations, the contributions of the steam locomotive are but footnotes to be occasionally studied in the textbooks of U.S. history.

And, at times, conveniently forgotten.

View # 1 of Union Pacific no. 9018 (4-12-2) after boiler explosion at Upland, Kansas, on October 20, 1948. Photo Courtesy Union Pacific Railroad.

View # 2 of Union Pacific no. 9018 (4-12-2) after boiler explosion caused by a dropped firebox crown sheet due to low water. Photo Courtesy Union Pacific Railroad.

View # 3 of Union Pacific no. 9018 (4-12-2) after boiler explosion that killed the three-man crew. Photo Courtesy Union Pacific Railroad.

1907 Luzon, New York, fatal train wreck caused by a boiler explosion.

Remains of Bessemer Railroad engine no. 507 that killed three following a boiler explosion at Hewitt Station, Pennsylvania, November 6, 1919.

Cause of the Hewitt Station boiler explosion: crown sheet failure and low water.

July 24, 1894, boiler explosion at Tucuman, Argentina Republic, launched boiler plate with dome 250 yards from accident scene. As a result of various equipment issues, only four of the country's 46 locomotives were operational at the time.

Tucuman boiler explosion resulted in eight fatalities.

A boiler explosion destroyed this 1850 steam engine revealing its internal components.

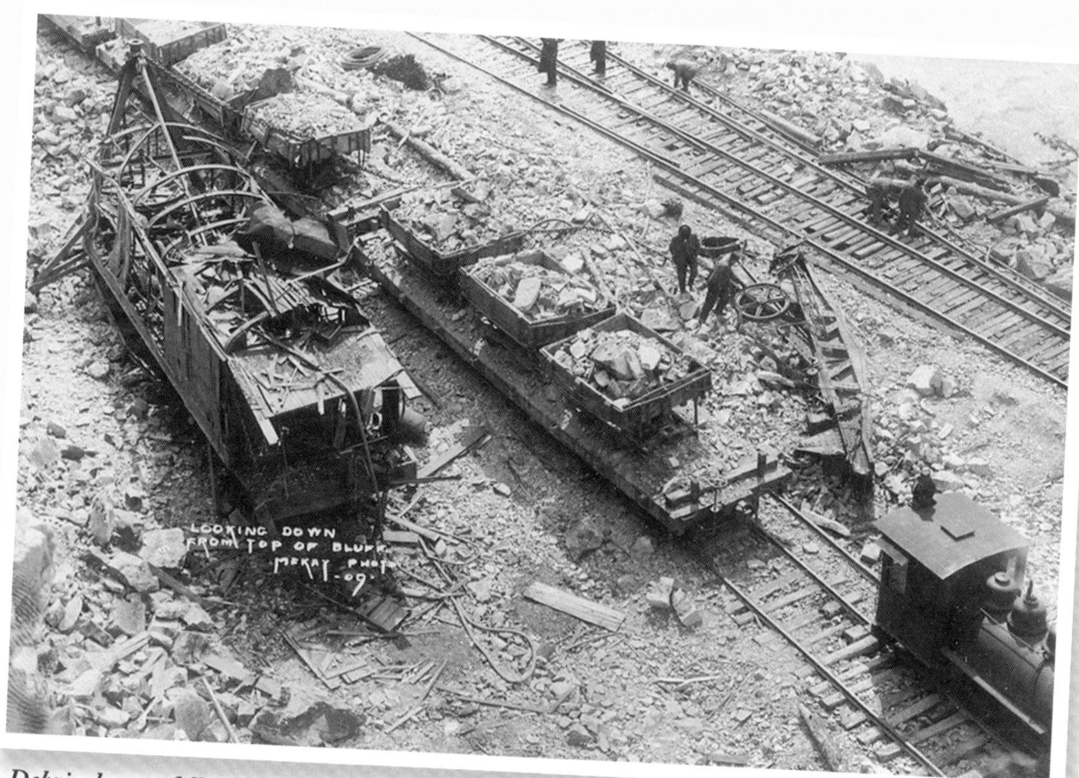

Debris cleanup following 1909 steam engine boiler explosion at Missoula, Montana.

OUR BROTHER

CHAPTER 16

FINAL JOURNEY OF THE *SULTANA*: North America's Deadliest Accident

(WARNING: GRAPHIC CONTENT)

Oh! To just think of the men that were there rushed into eternity without a moment's warning. I fear that but very few of them were prepared to meet their God. What a warning it should be to us all.

– *Erastus Winters*, Sultana *survivor, 1840–unknown*

It would be imprudent to discuss the dangers of pressure equipment without acknowledging one of the deadliest accidents ever recorded in North America.

A boiler explosion aboard the steamship *Sultana* occurred at two A.M., April 27, 1865, on the flooded Mississippi River just north of Memphis. It resulted in the deaths of an estimated 1,800 passengers and crew.

With a carrying capacity of 376, the *Sultana* was en route to Cairo, Illinois, with 2,300 passengers, many of whom were Union soldiers returning from confinement in Confederate prison camps and the ravages of the just-concluded Civil War.

By comparison to the more widely reported sinking of the Titanic (1,522 lives lost), the *Sultana* incident remains but a footnote to the Civil War, perhaps because it followed the assassination of Abraham Lincoln by less than two weeks.

Prior to its arrival in Memphis, the side-paddle steamer stopped at Vicksburg, Mississippi, where it experienced a bulging seam on the middle port boiler. A reluctant local boiler maker patched the seam without repairing the bulge, and the steamer was permitted to continue up the Mississippi River on its fateful journey.

Just after the *Sultana* took on coal near Memphis, it is believed one of the four high-pressure tubular boilers exploded, fragmenting the ship's entire center section. Scores of sleeping passengers were instantly killed or maimed. What followed resulted in the deaths and injury to hundreds upon hundreds of others, many of whom found themselves catapulted into the chilly Mississippi.

Today, nearly 150 years later, it is impossible for anyone to adequately describe this catastrophic tragedy.

But the voices of those fortunate to survive that hellish spring night must never be quieted. Their experiences reveal how pressure equipment malfunctions have a physical and emotional impact of vast proportion.

In addition to numerous letters written by victims to loved ones, there have been several excellent modern-day books that have sought to unravel the *Sultana* saga. However, perhaps the most important publication on this tragedy was penned 27 years after the accident.

Rare first edition of Chester D. Berry's **Loss of the Sultana and Reminiscences of Survivors.**

In 1892, *Sultana* survivor Chester D. Berry published a stirring and graphic account of the accident. *Loss of the* Sultana *and Reminiscences of Survivors* was a product of Berry's efforts to contact other survivors and catalog their first-hand experiences. Included are facts, records, and personal accounts—many of which illustrate a moving portrait of man's enduring will to survive.

Herewith an extraordinary look—directly from an exceptionally rare publication—at select passages featuring the haunting words of the emaciated Union prisoners who survived:

THE ATROCITIES OF PRISON LIFE

Most of us died more than a dozen living deaths while in prison, and looked more like candidates for the bone-yard than for anything else.

—J. S. Cook

Wood being so scarce I worked up one of my crutches to cook with before I was able to do without it. I next burned a part of the other one, and one night I failed to lie on my cane and some fellow stole it with which to cook his breakfast. Our prison was furnished with river water. The water passed through the city in pipes to a hydrant outside the prison where the stock came to drink. The people would wash there and then the water would pass down through the prison for us to drink and cook with, but still it was one of the purest articles we got.

—Perry S. Summerville

The food they gave us was corn cobs, all ground up and made into mush, and there wasn't near enough of that to keep the boys alive any length of time. Those that lived had to speculate by trading their brass buttons, boots, etc., with the guards. There were from one hundred to one hundred and fifty boys dying every day. A large wagon, drawn by four mules, was used in drawing out the dead. They were laid in as we pile cord wood and taken to the burying ground, generally putting fifty in a grave, and returning would bring mush in the same wagon, where worms that came from the dead could be seen crawling all over it; but we were starving, therefore we fought for it like hungry hogs.

—Joseph Stevens

We left in December for the Tennessee River. The ground was covered with ice and some of the boys had no shoes on—you could track them by the blood from their feet.

—Truman Smith

There was a small engagement took place there when our cavalry was surrounded and all taken prisoners, I being so unfortunate as to get shot through the arm near the shoulder. This was on the 30th day of October, 1864. They took me from Raccoon Ford to Florence, Ala., and there, for practice, the young rebel doctors cut off my arm; I think it could have been saved.

—Ogilvie E. Hamblin

We started for Cahaba, Ala., the prison known as "Castle Morgan." At this time my clothing consisted of shirt, drawers and one shoe.

—George F. Robinson

On the 30th of November, 1864, at the memorable battle of Franklin, Tenn., I was again taken prisoner and this time took a trip to Andersonville, that indescribable den of suffering, sorrow and death.

I will give some death rates that I gathered from official records as follows: Of 12,400 persons taken to the hospital, 76 per cent died.

In May, 1864, of 18,454 prisoners, 701 died, 23 per day.

In June, 1864, of 26,364 prisoners, 1,202 died; 40 per day.

In July, 1864, of 31,678 prisoners, 1,742 died; 56 per day.

In August, 1864, of 31,693 prisoners, 3,076 died, 99 per day.

On the 23rd of August, 1864, was the greatest mortality; 127 died, one for every eleven minutes. You will allow me to say that I call that treatment wholesale murder and that of the most cruel kind known to history.

—Samuel H. Raudebaugh

Oh! The long and dreary winter in prison; the suffering from cold, hunger, and the petty tyranny of cowards clothed with a little brief authority; the stench of rotten meat, of which we had not half enough to eat; the bitter, bitter feeling that our country had abandoned us to our fate, refusing to exchange because it would be exchanging able-bodied soldiers for us who were starved until we could be of no service.

Shall I tell of the march over ice and snow; the wading of deep streams from Nashville to Dixon, two miles below Cherokee, on the Memphis & Charleston railroad; the suffering from cold and exposure in the dead of winter, and from hunger; . . . of seeing a wagon load of corn on the ear, driven into the prison-corral and thrown out to us as though we were a lot of fattening hogs; of the number of dead we left on the ground next morning, killed by eating raw corn after a four days' fast; of my confinement, without food, for a day and night in a close, crowded box-car, in which fresh horse dung was half a foot deep . . . .

—J. Walter Elliott

BOARDING THE SULTANA IN MEMPHIS

We were . . . ordered to fall into line and march aboard the steamer "Sultana." When going on board my attention was attracted by the noise and work at the boilers going on at that time. We were marched to the hurricane deck and informed that this was to be our place of abode, but I thought different.

I went below and looked at the boilers, which were not very favorable to my mind. I went back to the boys, told them that we had better look for some other place and that I thought that there was danger; and if the boat should blow up and we were on that deck we would go higher than a kite.

—Wm. Boor

Last known photograph of* Sultana *and passengers in Helena, Arkansas: April 26, 1865, the day before its doomed voyage.

When about half of us were on board, the captain of the boat stopped us and said that he had enough, for he did not consider the boat safe enough to take so many as he had just had the boiler patched a few days before. The quartermaster, however, who had charge of us, swore that he was loading the boat and would put as many men on as he pleased.

The clerk replied that if we arrived safe at Cairo it would be the greatest trip ever made on the western waters, as there were more people on board than were ever carried on one boat on the Mississippi river. He stated that there were 2,400 soldiers, 100 citizen passengers, and crew of about eighty—in all over 2,500 . . . .

— A. C. Brown

There was no necessity of loading the "Sultana" so heavily, as the steamers "Pauline Carroll" and "Lady Gay" were at the landing coming up light, but the clerk and captain of the "Sultana" were part owners of the boat, and I understood at the time that they put up money to get the transportation of the soldiers, which the officers of the other boats, having no interest, would not do.

—Daniel McLeod

There were 2,000 soldiers and 200 passengers, besides some 700 hogsheads of sugar and I think about 30 or 40 mules and other freight.

—L. C. Morgan

. . . we joined some 500 privates from the stockade . . . All were begrimed and blackened by exposure, without a pretense of protection from summer's sun or winter's rain; all weak and lean from starvation; many, too feeble to take care of themselves, were literally encased in scales, beneath which were myriads of living vermin eating all vitality away. Two I saw doubled up and scarred all over, having been literally torn in pieces by the dogs, because they attempted to escape from the devil's domain.

—J. Walter Elliott

THE EXPLOSION

Myself and two comrades bunked together, just back of the left wheel house, on the middle deck. The first sensation I experienced was that of falling down through space, as probably many of you have felt when you had an attack of nightmare. I soon realized that it was no nightmare for we were immersed in the icy water of the river . . . .

—Jotham W. Maes

. . . [S]ome one asked me to help throw off the dead men, for it looked hard to see them burn. We threw off five or six. One poor fellow was pinned down by the wreck and begged some one to help him out. I tried to but the timbers were so heavy that I could not get him loose and so I had to let him burn to death.

—L. C. Morgan

I had taken off all my clothes except my drawers and vest; in the latter was a diary and pictures of my wife and girls; these I saved.

—George A. Clarkson

Men were scalded and burned, some with legs and arms off, and it seemed as if some were coming out of the fire and from under the boiler, and many of them jumping into the river and drowning by squads.

—Nathan S. Williams

When the explosion occurred it threw the boiler out of its bed, ascending and tearing its way through both cabin and hurricane decks. Those immediately over the boiler were thrown in every direction, some of them being thrown directly up and falling into the fiery chasm below, while those upon the side of the boat, like myself, were thrown directly out and away from the boat. The first I realized after the explosion I found myself about 300 feet from the boat shrouded in total darkness and in what appeared to be an ocean of water.

Now I was alone, cold and tired. I began to look around for some support, which I found in the shape of an empty candle box which answered the purpose very well. This box I still had in my possession when picked up by a skiff eighteen miles below where the accident took place.

— John H. Kochenderfer

When the opportunity presented itself I took off my blouse, hat and shoes, keeping on all my underclothing, and took an ambrotype likeness of my wife and boy, out of my blouse pocket and put it in my pants pocket so that if I was lost and ever found it would be the means of identifying me.

—Benjamin F. Johnston

I saw boys start out to swim with all their clothes on, even their overcoats and shoes, but they did not go far before they sank.

—William H. Peacock

"Captain, will you please help me?"

I turned in the direction of the voice so polite, so cool and calm amid this confusion. There, on the head of the last cot on this side of the breach, which was covered with pieces of the wreck, sat a man, bruised, cut, scalded in various places, both ankles broken and bones protruding. With his suspenders he had improvised tourniquets for both legs, to prevent bleeding to death.

"I am powerless to help you; I can't swim," I replied. But he answered, "Throw me in the river is all I ask; I shall burn to death here."

I worked and toiled to my very utmost to assist others, until all was done that I could do. Then the thought occurred to me that it was my duty to make an effort to save myself. I saw two Kentuckians meet, each lamenting that he could not swim. "Then let us die together," said one. "Well," replied the other, and, embraced in each other's arms, they leaped, sank, and the muddy waters closed over them. I saw others, blinded by the explosion, leap into the fire and die.

— J. Walter Elliott

. . . [O]verboard I started. Before I reached the water something was thrown over that hit me and down I went under the water. As I came up a drowning man caught me around the neck with a death grip, and under we went—the second time for me. As we sank I strangled. I now passed through the same experience that only a drowning person or those about to drown undergo. In those few seconds of time my whole life, from my childhood down to that terrible moment, passed before me like a panorama with perfect distinctness. As we came to the surface I freed myself from his deadly grasp and struck out for myself.

—A. C. Brown

As I stopped to take a hurried glance around me I heard some one near me exclaim, "For God's sake some one help me get this man out." I turned and saw a lieutenant of a Kentucky regiment. He was a very large man and was called "Big Kentuck." He had found a man that was held fast by both feet, a large piece of the wreck having fallen across them. I took hold and helped the lieutenant but we could not release him and he was soon roasted by the intense heat.

— M. C. White

At the time her boilers exploded I was lying sound asleep on the lower deck, just back of the rear hatchway to the hold. I was not long in waking up, for I was nearly buried with dead and wounded comrades, legs, arms, heads, and all parts of human bodies, and fragments of the wrecked upper decks.

And now occurred the hardest task of my life. The boat was on fire and the wounded begged us to throw them overboard, choosing to drown instead of being roasted to death. While our hearts went out in sympathy for our suffering and dying comrades we performed our sad but solemn duty. When . . . the last shadow of hope had expired . . . we proceeded to perform carefully, but hurriedly, the most heart rending task that human beings could be called upon to perform—that of throwing overboard into the jaws of certain death by drowning those comrades who were unable on account of broken bones and limbs to help themselves. Some were so badly scalded by the hot water and steam from the exploded boiler

Rare antique woodcut engraving of the* Sultana *explosion published in* Harper's Weekly, *May 20, 1865.

that the flesh was falling from their bones . . .

Readers of this narrative, do you not think that this was a hard task for us to perform? If not, just hearken to this a moment; listen to the heartfelt prayers of those suffering and wounded comrades and hear their dying requests as they commended their wives, children, fathers, mothers, sisters and brothers to God's kind care and keeping, and hear them thanking us for our kindness to them, notwithstanding the pain they were suffering. They fully realized the fact that their last day, hour and even last minute to live had come; and then to hear the gurgling sounds, and dying groans and see them writhing in the water, and finally see them sink to rise no more until the morning when all shall come forth. Was this not heart rending to us? My heart, even now, after twenty-seven years, nearly stands still while I write this sad story.

—Commodore Smith

The steam and ashes smothered us so we could scarcely breathe. When I came to my senses I rushed for the stern entrance, falling several times before I reached the fresh air. Now hundreds of men came rushing out to get breath. Jamming and crowding commenced. Those crippled were trampled on.

—Albert W. King

Such hissing of steam, the crash of the different decks as they came together with the tons of living freight, the falling of the massive smoke stacks, the death-cry of strong-hearted men caught in every conceivable manner, the red-tongued flames bursting up through the mass of humanity and driving to death's door those who were fortunate enough to live through worse than a dozen deaths in that "damnable death pen" at Andersonville. We had faced death day by day while incarcerated there, but this was far more appalling than any scene through which we had passed.

But oh, what a change in one short moment! Comrades imploring each other for assistance that they might escape from the burning deck; officers giving orders for the safety of their men; women shrieking for help; horses neighing; mules kicking and making the terrible scene hideous with their awful brays of distress. These are a few of the many scenes and sounds that greeted my sight and came to my ear.

Can I ever forget the scene? Not while my senses remain. Masses of drowning men clinging together until they were borne down by their own weight to rise no more alive. Their poor, pinched, and ghastly faces are indelibly engraved on my memory.

Those noble men who had faced battle in all its fury; who had not flinched when the word "forward" came, even though in the face of the cannon or screaming shell; had faced worse than death at Andersonville; standing there on the bow of that burning boat wringing their hands, rushing to and fro begging and imploring their comrades to assist them that their lives might be saved to their dear ones!

—A. A. Jones

One of these ladies, with more than ordinary courage, when the flames at last drove all the men from the boat, seeing them fighting like demons in the water in the mad endeavor to save their lives, actually destroying each other and themselves by their wild actions, talked to them, urging them to be men, and finally succeeded in getting them quieted down, clinging to the ropes and chains that hung over the bow of the boat. The flames now began to lap around her with their fiery tongues. The men pleaded and urged her to jump into the water and thus save herself, but she refused, saying: "I might lose my presence of mind and be the means of the death of some of you." And so, rather than run the risk of becoming the cause of the death of a single person, she folded her arms quietly over her bosom and burned, a voluntary martyr to the men she had so lately quieted.

—Chester D. Berry (book editor)

I was thinking whether to burn or drown, when a woman with a little babe about two months old came to me crying for help. I told her it was every one for himself. I saw that she had on a life preserver but it was buckled down too low. I stepped up to her and was going to unbuckle it, when she said, "Soldier, don't take that off from me." I said, "it must be up under your arms." I placed it there, and took her by the hand and she jumped into the water. She thanked me and said, "may the Lord bless you." She lost her husband, baby, father, and mother there.

— Nathaniel Foglesong

I went to the bow of the boat to see what had become of the man that was killed. He was still there but all of his clothing was torn off him by the men running over his body.

We both went to the bow of the boat to jump overboard, but there were too many men in the water, the water being covered with men's heads, all of them begging for something to be thrown to them on which they might escape. I believe I saw 150 or 200 men sink at once near the bow of the boat.

The 102d Ohio Volunteer Infantry had one hundred and five (105) men on the boat and only thirty-two were saved. Out of fourteen men in Company H only three were saved.

—Simeon D. Chelf

I was on the second deck, my partner's name was Joseph Test, from Dayton, Ohio. A piece of timber ran through his body, killing him almost instantly . . .

On board the boat was a pet alligator. He was kept in the wheel-house. It was a curiosity for us to see such a large one. We would punch him with sticks to see him open his mouth, but the boatmen got tired of this and put him in the closet under the stairway. When I came down stairs every loose board, door, window and shutter was taken to swim on, and the fire was getting very hot. I thought of the box that contained the alligator, so I got it out of the closet and took him out and ran the bayonet through him three times. While I was doing this a man came to me and said the box would do for he and I both to get out on. My intention was to share it with him, but I did not speak and I do not know what became of him. I took off all my clothing except my drawers, drew the box to the end of the boat, threw it overboard and jumped after it but missed it and went down somewhere in the mighty deep . . .

There were hundreds of men in the water and they would reach for anything they could see. When a man would get close enough I would kick him off, then turn quick as I could and kick someone else to keep them from getting hold of me. They would call out "don't kick, for I am drowning," but if they had got hold of me we would both have drowned.

—William Lucenbeal

There were so many people in the water you could almost walk over their heads. I got a shutter about three feet square, and at this time I found Joe Moss. He begged me to let him have the shutter as he could not swim. I threw it into the river and told him to follow it, which he did; I never saw him again.

—Ben C. Davis

WATER RESCUE

As I swam away I heard someone coughing and swam towards him. As I came near he kept swimming away. I called him and asked what regiment he belonged to. He asked what I wanted to know for. I told him I would write to his people in case he drowned and I should get out. He said I must not come any closer, and we made a bargain that if one should die and the other get ashore the survivor should write the parents and let them know. We kept swimming till near daylight . . . .

—Truman Smith

We were going along fairly well when a drowning man seized my left leg. I tried to kick him loose but failing I let go the raft and tried to force him off but could not, and was obliged to drag that dead weight until we reached Memphis. We were helped out of the water just above the wharf by citizens, and the last I can recollect was they were trying to pry the dead man's grip loose from my leg.

—M. H. Sprinkle

I also found out that I was scalded about the face, and every now and then plunged my face into the water to cool it off. I made no effort to swim ashore, as I knew the river had overflowed its banks, and I did not relish the idea of climbing a tree to get out of the water when I had nothing on but my shirt. That would be very unpleasant.

Then I saw the boatmen row towards me and they pulled me in the boat. I now found I was badly scalded on my left side and back. After picking out of the river a few more of my unfortunate comrades the boat rowed for the shore and we were landed and taken to a convalescent camp, where I continued to suffer the most excruciating pain. I ran up and down in the cool air to relieve the pain. I felt easy going against the wind, but returning it was excruciating.

—C. S. Schmutz

It was a mule that saved my life and a dead one at that. I was almost a goner, when I saw a dark object in the water and made for it, and it was a dead mule, one that was blown off the boat. He was dead but not quite cold. I crawled up on him and was there when I was picked up at Fort Pickens three miles below Memphis.

—George F. Robinson

Comrade John Cornwell of my company and regiment and myself swam together, but he was easily discouraged. After awhile he called out to me that he could hold out no longer, but I cheered him up, urging him to try a little longer, telling him that I knew he was just as able to get out as I and that I was not going to give up. He tried awhile longer and then cried out again that it was no use, he must sink. I urged him to hold on, but after we had gone about two miles he called a third time and sank immediately, and I saw him no more.

—L. W. McCrory

The water was full of struggling and drowning people. I heard a lady crying for help, asking her husband to rescue her. She was holding to a rope attached to a mule that had got overboard. I also saw the husband, with a little child on his back, struggling in the water for a moment, then sinking. The lady cried out, "My husband and baby are gone!" A comrade who had his limb crushed in the explosion by a door blown from the boat had the lady get on this door, through which means she was rescued.

—C. J. Lahue

Owing to the necessity of constant motion, without rest to any part of the body, being reduced to a mere skeleton through being confined in rebel prisons was in my favor, as I could never have survived that awful disaster had I weighed as much as I did before my prison experience. My weight now was eighty pounds. When I was captured I weighed 175 pounds.

—Epenetus W. McIntosh

I stood still and watched for a while, then began wandering around to other parts of the boat when I came across one man who was weeping bitterly and wringing his hands as if in terrible agony, continually crying, "O dear, O dear." I supposed the poor fellow was seriously hurt. My sympathies were roused at once. Approaching him, I took him by the shoulder and asked where he was hurt. "I'm not hurt at all," he said, "but I can't swim, I've got to drown, O dear." I bade him be quiet, then showing him my little board I said to him, "there, do you see that; now you go to that pile of broken deck and get you one like it, and when you jump into the water put it under your chin and you can't drown." "But I did get one," said he "and some one snatched it away from me." "Well then, get another," said I. "I did," said he, "and they took that away from me." "Well, then," said I "get another." "Why," said he, "what would be the use, they would take it from me. O dear, I tell you there is no use; I've got to drown, I can't swim." By this time I was thoroughly disgusted, and giving him a shove, I said, "drown then you fool."

I want to say to you, gentle reader, I have been sorry all these years for that very act.

— Chester D. Berry (book editor)

I can well remember seeing the captain putting life preservers on his wife and little girl and letting them overboard. The girl's life preserver slipped too far down for she was found (drowned) floating with her feet upwards. His wife was saved and the captain lost his life in trying to save others. We had a number of mules aboard the boat and some of the boys hung on to their tails while they swam to shore.

—Joseph Stevens

The dock was covered with ladies belonging to the Christian and Sanitary Commissions, who gave us each a pair of drawers and a shirt. I started up town, but at the first block I came to there was a great crowd and they wanted to know if I was on the boat. I said yes, and they gave me a suit of clothes and thirteen dollars in money.

—Truman Smith

I soon came in contact with a log upon which I crawled, and where I remained until about nine o'clock the next day, when I was taken off by the steamer "Pocahontas." While upon this log I saw a man reach an island who was pulled out by two of his comrades. I do not believe there was a particle of skin upon his entire body. He had been badly scalded and it had all come off. His comrades were doing their best to keep the buffalo gnats off him. What ever became of the poor fellow I never knew, but presume that he died in a short time.

. . . the first man I came across on the "Pocahontas" . . . was dishing out hot sling unsparingly to the boys. I took a big drink but it was not enough, so I went up to the bar of the boat and called for brandy. The bartender set down a bottle and a small glass, but I called for a large one. He then set down a big beer tumbler. I filled this brimming full and drank it, then offered to pay

for it, but he refused to take pay, saying "it is free to 'Sultana' survivors."

— L. W. McCrory

They poured whiskey down me, rolled and rubbed me, and finally brought me back to life. I was like the new born babe, not a raveling of clothing upon me.

—Hosea C. Aldrich

At the time the "Sultana" blew up I was thrown from the boiler deck and very badly hurt, but was fortunate enough, with three unknown comrades, to get hold of a bale of hay, upon which we floated till nearly opposite the city of Memphis, where we were picked up by a boat.

—Wm. Barnes

I think there were about fifteen or sixteen of us that had stuck to the plank. But now a new danger had seized me, as some one grabbed me by the right foot and it seemed as though it was in a vise; try as I would, I could not shake him off. I gripped the plank with all the strength that I had, and then I got my left foot between his hand and my foot and while holding on to the plank with both hands I pried him loose with my left foot, he taking my sock along with him, but he is welcome to the sock; he sank out of sight and I saw him no more.

—James K. Brady

After getting back on the old wreck I met Thomas Pangle of my company and saw the bodies of three men that were burned beyond recognition, and helped to pull a man up on the boat; he was one of the engineers. His nose was torn off, all except a small particle of skin, and he died before he was taken to land.

The two men that rescued us brought ashore the bodies of two dead women, mother and daughter, who were of a family of about eight persons, all of whom were drowned except a grown son who was frantic with grief at the sight of his dead mother and sister.

I found Jarson M. Elliott of our company on the boat. He was scalded all over and unable to help himself, but was perfectly composed and bore his suffering with great fortitude. He had his army badge which he requested me to give to his parents. He died that night at Gayoso Hospital, Memphis, Tenn.

—Robert N. Hamilton

RECOVERY

I was placed in a ward with quite a number who were severely scalded, or otherwise badly injured, and such misery and intense suffering as I witnessed, while there is beyond my power to describe. The agonizing cries and groans of the burned and scalded were heartrending and almost unendurable, but in most cases the suffering was of short duration as the most of them were relieved by death in a few hours.

—William Fies

. . . [M]y mother received official notice from Washington that her son was killed upon the

"Sultana;" and my name stands today upon the Michigan Adjutant General's Report for 1865 as killed by the explosion of the steamer "Sultana."

—Chester D. Berry (book editor)

RETURNING HOME

I stayed in Memphis about two weeks and met my friend Everman, who was very glad to see me. We were afraid to try the boats again and waited for the train to go North.

—Samuel C. Haines

No adequate cause for the explosion has ever been ascertained. The steamer was running at her proper speed (nine or ten miles an hour). This casualty transpired in time of intense excitement and never had the attention it ought to have had, following closely as it did the assassination of President Lincoln and the close of the war. Death and destruction had been in the land for four years, and nearly 400,000 had already given up their lives in defense of the national flag, that it might wave over a free country.

—Isaac Van Nuys

The people of the town called a meeting in a new hotel, which was not completed inside yet. That evening the local speakers of the town made several patriotic speeches to us, but what was the nicest thing of all there were about forty ladies, dressed in red, white and blue, that sang several patriotic songs; among the rest they sang "Welcome home, dear brothers," and it seemed that we were. Ever since that time I have had a warm place in my heart for the people of Mattoon and surrounding country, also for the people of Cairo, Ill. But all things have an end, and so at one o'clock we started for Columbus, the capital of the great and glorious old state of Ohio. In due time we arrived; but oh, what a change, instead of being treated like lords, as we were in Illinois, we were treated more like so many dogs than human beings. . . I came home and went around to see my friends and neighbors, but when I went around it seemed as though everybody was gone or dead.

—James K. Brady

In due time we arrived at Cairo, and after getting transportation from the quartermaster's department, were sent to Columbus, some to "Camp Chase," the injured ones to "Treplar Hospital," where right in sight of the capitol of our own glorious state of Ohio we were treated more like brutes than soldiers, and were almost starved to death by some inhuman, dishonest scoundrel, in the employ of the government.

—William Fies

After being at the hospital a few days, and not being injured, I made my escape, determining to reach home as soon as possible. The first boat that came along was the "St. Patrick," a handsome steamer plying between Cincinnati and Memphis. Like a burnt child dreading the fire, I dreaded getting on a steamboat for fear of another explosion. Adopting what I supposed was the safest plan, I crawled into the yawl hanging over the stern of the boat (as all sidewheel packets have) and never left my quarters until I arrived at the wharf at Evansville. It rained most all the way up, but I stuck it through. Every time the boat would escape steam or blow the whistle I prepared to jump, supposing an explosion was about to take place.

—Wm. A. McFarland

LIFE AFTER SULTANA

I believe that some enemy of our Union had a hand in crowding so many of us on the boat, and that he knew when that southern sugar was taken off that the rest of the cargo and the boat would meet the fate that followed. I believe that some ally of Jeff Davis put a torpedo in the coal, while we were at Memphis, where it would go into the furnace for the first fire that would be built after leaving Memphis, with the intent to destroy the boat and its mass of human heroes on their way home. I can say that in May, 1888, a man in the south, William C. Streeter, St. Louis, Mo., said that he knew the man, Charles Dale, who said he chiseled a hole in a large chunk of coal, put the torpedo therein which did the deadly work, carried it with his own hands and laid it where it must soon go into the furnace.

—Samuel H. Raudebaugh

I have seen death's carnival in the yellow-fever and the cholera-stricken city, on the ensanguined field, in hospital and prison, and on the rail; I have, with wife and children clinging in terror to my knees, wrestled with the midnight cyclone; but the most horrible of all were the sights and sounds of that hour. The sight of 2,000 ghostly, pallid faces upturned in the chilling waters of the Mississippi, as I looked down on them from the boat, is a picture that haunts me in my dreams.

— J. Walter Elliott

One of 700 survivors of the *Sultana* catastrophe was Corporal Erastus Winters of Company K of the 50th Ohio Infantry. In this rare 1865 letter to his sister Phebe Winters, the Union soldier reports on his condition from a Memphis hospital less than two weeks after the accident. In addition to surviving the *Sultana* sinking, Corporal Winters also fought in the Battle of Perryville (Ky.) and the Battle of Franklin, Tenn. In 1905, he would publish *THE CIVIL WAR MEMOIRS OF ERASTUS WINTERS* chronicling his adventures as a federal trooper and Confederate prisoner of war.

Erastus Winters.

Western Branches of the
U. S. C. COMMISSION.
CHICAGO: J. V. Farwell, Chairman,
B. F. Jacobs, Secretary.
MILWAUKEE: W. S. Carter, Chairman, D. W. Perkins, Secretary.
PEORIA: A. G. Tyng, Chairman,
Wm. Reynolds, Secretary.
ST. LOUIS: J. W. McIntyre, Chairman, J. H. Parsons, Corresponding Secretary.

THE U.S. CHRISTIAN COMMISSION
Sends this as the soldier's messenger to his home. Let it hasten to those who wait for tidings.

U. S. Christian Commission Rooms.

Memphis Tenn May 20th 1865.

Dear Sister
I embrace the present opportunity
of writing you a few lines to let you
know I am Still alive and well I
am Still in Adams General Hospital
No 8 at Memphis Tennessee but I got a
pass this Morning and came into the
Christian Commission rooms and I thought
I would write you a few lines to let
you know that the Lord Still Sees fit
to Spare my poor unprofitable life
I wrote father a letter a few days ago
So I Suppose you have heard from me
by this time I am getting along fine
my burns are nearly all healed up and
I think I Shall Start up the River in
a few days Well I dont care how
Soon for I am geting very uneasy
about you folks at home because I
have not heard a Word from you Since
your letter of the 6 of March I am
very fearful that Something has happened
but I think if the Lord is Willing
I Shall be home in a Short time

We are having the finest of
Weather here now and I should enjoy
a trip up the river now very much
provided I did not get Blown up again
Which I pray the Lord I may not
I want to get home in time for
Strawberys if I can Strawberries are Ripe
here now but they dont do me
any good as I have no money
Well as I have nothing of importance
to write to you at this time I will
Close hoping I may meet you all
in a few days again and Will
I am dear Sister yours affectionate
and long absent Brother

Erastus Winters

Miss Phebe Winters

Dear Sister

I embrace the present opportunity of writing you a few lines to let you know I am still alive and well. I am still in Adams General Hospital No. 8 at Memphis Tennessee but I got a pass this morning and came into the Christian Commission rooms and I thought I would write you a few lines to let you know that the Lord still sees fit to spare my poor unprofitable life. I wrote father a letter a few days ago. So I suppose you have heard from me by this time. I am getting along fine. My burns are mainly all healed up and I think I shall start up the river in a few days. Well I don't care how soon for. I am getting very uneasy about you folks at home because I have not heard a word from you since your letter of the 6 of March. I am very fearful that something has happened but I think if the Lord is willing I shall be home in a short time. We are having the finest of weather here now and I should enjoy a trip up the river now very much provided I did not get blown up again which I pray the Lord I may not. I want to get home in time for strawberries if I can. Strawberries are ripe here now but they don't do me any good as I have no money. Well as I have nothing of importance to write to you at this time. I will close hoping I may meet you all in a few days alive and well. I am dear sister your affectionate and long absent brother

Erastus Winters

ABRIDGED U.S. STEAMBOAT BOILER EXPLOSION TIMELINE

NOTE: No comprehensive records were kept by either the U.S. government or industry during many of the following years.

1816

- Steamboat *Washington* (14 dead), 3 July, maiden voyage
- Steamboat *Maid of New Orleans* (6 dead)

1825

- Steamboat *Teche* (60 dead), 5 May

1836

- Steamboat *Ben Franklin* ("great loss of life"), 13 March
- Steamboat *Motto* (11 dead), August

1837

- Steamboat *Chariton* (3 dead), 27 July
- Steamboat *Dubuque* (16 dead), 15 August

Engraving depicting explosion onboard the* Washington, *the first explosion on western waters.
Engraving courtesy of Williams Research Center, Historic New Orleans Collection.

1838

- Congress establishes U.S. Steamboat Inspection Service, January
- Steamboat *Oronoko* (70 dead), 15 April
- Steamboat *Moselle* (230 dead), 26 April

1839

- Steamboat *George Collie,* (26 dead, 45 scalded), 26 April

1841

- Steamboat *Louisiana* (23 dead), August
- Steamboat *Erie (*175 dead), 9 August

1843

- Steamboat *Clipper* ("grave loss of life"), 19 September

1844

- Steamboat *Lucy Walker* (18 dead), 25 October

Engraving illustrating* Moselle *boiler explosion. Courtesy of Kentucky Historical Society Collections.

1845

- Steamboat *H. Kinney* (7 dead), 28 March
- Steamboat *Big Hatchie* (35 dead), 25 July
- Steamboat *Denizen* (1 dead), 30 November

1846

- Steamboat *Concordia* (28 dead), 16 September

1847

- Steamboat *Tuskaloosa* (12 dead), 29 January
- *Steamboat *Simon Kenton* (1 dead), 28 April
- Steamboat *Island Packet* (2 dead), 9 December
- Steamboat *Westwood* (12 dead), 19 December
- Steamboat *Phoenix* (240 dead), November

**Burst steam lines*

1848

- Steamboat *Blue Ridge* (13 dead), January
- Steamboat *Planter* (5 dead), 5 January
- Steamboat *H. Kinney* (7 dead), 28 May
- Steamboat *Olive* ("with loss of life"), 2 September

1849

- Steamboat *Virginia* (10 dead), 31 March
- Steamboat *Emily* (7 dead), 27 May
- Steamboat *Louisiana* (86 dead), 15 November

1850

- Steamboat *Hope* (5 dead), 27 February
- Steamboat *Ironton* (3 dead), May
- Steamboat *Kate Fleming* (11 dead, 6 injured), 5 October
- Steamboat *Financier* (2 dead), 12 October
- Steamboat *Sagamore* (50 dead), 29 October
- Steamboat *G.P. Griffin* (945 dead), 17 June

Engraving courtesy of Williams Research Center, Historic New Orleans Collection.

1851

- *Steamboat *New World* (7 dead), January
- Steamboat *Brilliant* (3 dead), 20 January (boiler flue collapse)
- Steamboat *St. Louis* (20 dead), 20 February
- Steamboat *Oregon*, (21 dead), 2 March
- Steamboat *Gretna* (3 dead), 17 June
- Steamboat *Brilliant* (47 dead or injured), 31 September

1852

- Steamboat *Pitser Miller* (3 dead), 23 January
- Steamboat *Redstone* (14 dead), 2 April
- Steamboat *Glencoe* (40+ dead), 4 April
- Steamboat *Saluda* (27 dead), 9 April
- Steamboat *St. James* (30 dead), 5 July
- Steamboat *Buckeye Belle* (20 dead, 14 injured), 12 November
- Steamboat *Geneva* ("many dead"), 2 December

1853

- Steamboat *R. K. Page* (3 dead), 22 March
- Steamboat *Farmer* (32 dead), 23 March

1854

- Steamboat *Helen Hensley* (2 dead), 19 January
- Steamboat *Georgia* (25 dead), 28 January
- Steamboat *Kate Kearney* (6 dead, many injured), 16 February
- Steamboat *Secretary* (16 dead, 30 scalded), 15 April
- Steamboat *Beardstown* (1 dead), 8 June
- Steamboat *Timour* (19 dead), 20 August
- Steamboat *Daniel Pratt* (3 dead), 6 October

1855

- Steamboat *Heroine* (3 dead), 13 March
- Steamboat *Sampson* (2 dead), 31 May
- Steamboat *Lexington* (30 dead), 30 June

1856

- Steamboat *Northern Indiana* (320 dead), 17 July
- Steamboat *Sarah* (2 dead), 21 February

1857

- Steamboat *Major A. Harris* (1 dead), 6 February
- Steamboat *Forest Rose* (14 dead), 25 March
- Steamboat *Montreal* (250 dead), 27 June

**Burst steam lines*

1858

- Steamboat *Post Boy* (3 dead), 10 January
- Steamboat *Fanny Fern* (20 dead), 20 January
- Steamboat *Colonel Crossman* (14 dead), 2 February
- Steamboat *Pennsylvania* (250 dead) 13 June
- Steamboat *Titania* (1 dead), 13 October

1859

- Steamboat *Princess* (70 dead), 27 February
- Steamboat *St. Nicholas* (60 dead), 24 April
- Steamboat *Michigan* (2 dead), 9 December

1860

- Steamboat *John C. Calhoun* (6 dead), 28 April
- Steamboat *H.R.W. Hill* (39 dead), 31 October

1861

- Steamboat *Corrine* (15 dead), 28 February

THE EXPLOSION OF THE STEAMER "PRINCESS" AT CONRAD'S POINT, ON THE MISSISSIPPI.

Engraving courtesy of Williams Research Center, Historic New Orleans Collection.

1862

- Steamboat *Ceres* (12 dead), 9 October

1863

- Steamboat *Adda Hancock* (49 dead), 27 April

1864

- Steamboat *Frank Steele* (2 dead), 2 June
- Steamboat *Washoe* (16 dead, 51 injured), 5 September
- Steamboat *J.C. Irwin* (11 dead), 17 October

1865

- Steamboat *Eclipse* (27 dead), 27 January
- Steamboat *Sultana,* (1,800+ dead), 27 April
- Steamboat *Burd Levi*, (5 dead), 19 May
- Steamboat *Joseph Pierce* (12 dead), 31 July
- Steamboat *Nimrod* (5 dead), 22 September
- Steamboat *Yosemite* (42 dead, 30 injured), 12 October (steam drum)
- Steamboat *St. John* (9 killed, 15 injured), 29 October
- Steamboat *De Soto* (11 dead), 9 December

1866

- Steamboat *W.R. Carter* (125 dead), 9 February
- Steamboat *R.J. Lockwood* (25 dead), 4 March
- Steamboat *City of Memphis* (11 dead), 31 May
- Steamboat *General Lytle* (35 dead), 6 August
- Steamboat *Julia* (5 dead, 11 injured), 30 September

1867

- Steamboat *David White* ("large loss of life"), 17 February
- Steamboat *Lansing* ("considerable loss of life"), 13 May
- Steamboat *Cuba* (7 dead), 11 November

1868

- Steamboat *Harry Dean* (5 dead), 4 January
- Steamboat *Magnolia* (35 dead), 18 March
- Steamboat *Seabird*, (100 dead), 10 April
- *Steamboat *Carrie Williams* (1 dead, 1 injured), 26 May
- Steamboat *Glide* (15 dead), December
- Steamboat *Volant* (1 dead), 20 December

**Burst steam lines*

EXPLOSION OF THE BOILER ON BOARD THE "ST. JOHN," OCTOBER 29, 1865—SCENE IN THE MAIN SALOON.—[SKETCHED BY JOHN P. NEWELL.]

EXPLOSION ON THE "ST. JOHN."

WE give on this page three illustrations relating to the recent distressing catastrophe on board the steamer *St. John*. This boat is one of the finest river boats in the world, and plies between Albany and New York. On the morning of Sunday the 29th ult., as she had arrived opposite Thirtieth Street on the way to her landing at Canal Street in this city, her boiler burst, and thirty tuns of scalding water with an immense volume of steam rushed through the huge fissure in the boiler into the state-room hall, carrying destruction in its track. Almost instantly nine persons were killed. The state-rooms in the immediate vicinity of the boiler were completely destroyed, and their inmates were killed. The boiler on one side being empty the vessel careened over on her other side, and the pool of scalding water followed the motion of the boat toward the opposite state-rooms. At the moment of the accident many fell victims to the destroying current by opening the doors of their state-rooms to discover what was the matter, and were stifled and scalded by the steam.

The scene in the main saloon, as presented by our artist, was heart-rending. Frightened passengers leaped barefoot into the hot water. Poor, struggling creatures lay helpless and bruised, tossing in unutterable agony.

Among the fatally injured were Captain F. J. LYON and his wife, MARY IMOGENE LYON. When the explosion took place Mr. and Mrs. LYON, both undressed, rushed into the state-room hall. At that time the hall was full of steam, and both were fearfully scalded. They had been married only three days, and were returning from their wedding-trip to Albany only to be buried in the Church of St. Luke's, where they had been married. Besides those killed fifteen were severely injured. JOSEPH LAMBERT, fireman, rushed overboard when the explosion took place and was drowned. Mr. REYNOLDS and his child were providentially saved, having left their state-room just before the accident.

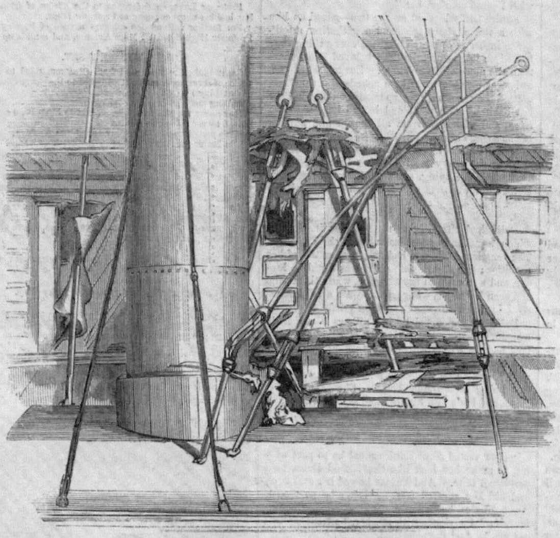

EXTERIOR VIEW OF THE STATE-ROOMS OVER THE BOILER.—[SKETCHED BY JOHN P. NEWELL.]

The wife and mother, however, was left behind, and was among the killed.

The Hoboken ferry-boat *Morristown*, seeing the signal of distress flying, went to the assistance of the unfortunate passengers, and succeeded in taking a large number of them off the steamer and landing them safely; while those on shore did all in their power to alleviate the sufferings of the scalded, and tried to make the others as comfortable as circumstances would permit.

A coroner's investigation has been held, but no satisfactory result can be attained from the evidence presented.

Mr. NORMAN WIARD of 46 Pine Street, New York, has written a paper upon boiler explosions, which is of great interest in connection with this disaster. These explosions, he says, are *not* due,

1st, to the direct pressure of steam, for it is not the weakest part that gives way; nor,

2d, to the decomposition of steam, giving it the quality and character of an explosive gas; nor,

3d, to the overheating of the boiler-plates; nor,

4th, to the increased pressure of steam from injecting water up on heated plates after the water has been low for a previous time, since not enough extra steam is produced in this way; nor,

5th, to the projection of a large amount of "solid water" with momentum against the shell of the boiler, when the pressure is relieved suddenly from the surface; nor,

6th, to the mysterious action of electricity; nor,

7th, to a new sudden increase of steam from the overheating of steam, plates, and stays, the surplus heat being communicated to the water.

Mr. WIARD attributes these explosions to the unequal expansion of the boiler plates, they being heated in one part more than in another, while there is the ordinary or less than the ordinary steam pressure. Successive unequal expansions of this sort in some particular part permanently weaken the iron in that part, and hence the explosion. Mr. WIARD considers this to have been the cause of the explosion on the *St. John*.

THE HUDSON RIVER STEAMER "ST. JOHN."

1869

- Steamboat *Cumberland* (18 dead), 14 August
- Steamboat *Phantom* (5 dead), 15 September

1870

- Steamboat *Silver Spray* (9 dead), 10 April
- Steamboat *Right Of Way* (9 dead), 27 July

1871

- Steamboat *Judge Wheeler* (3 dead), February
- * Staten Island Ferryboat (125 dead) 30 July
- Steamboat *Sam J. Hale* (6 dead), April

1872

- Steamboat *Oceanus* (34 dead), 11 April

1873

- Steamboat *George C. Wolff* (12 dead, 15 injured), 22 August

1875

- Steamboat *Hugh Martin* (1 dead), 14 August

1876

- Steamboat *Pat Cleburne* (14 dead), 17 May

1877

- Steamboat *Comfort* (10 dead), 3 October

1878

- Steamboat *Sandy Fashion* (2 dead), 6 April
- Steamboat *Phil Morgan* (3 dead), 28 April
- Steamboat *Sherley-Bell* (1 dead), 17 September

1879

- Steamboat *L.C. McCormick* (1 dead), 16 February

1880

- Steamboat *Bonnie Lee* (9 dead), 9 August

1881

- Steamboat *Phaeton* (4 dead), 28 June

1882

- Steamboat *George Washington* (17 dead, 10 injured), 14 January

**Burst steam lines*

***Engraving published in the September 16, 1871,* Harper's Weekly *of a boiler explosion aboard the* Judge Wheeler.** Courtesy of Williams Research Center, Historic New Orleans Collection.

1885

- Steamboat *Mark Twain* (6 dead), 27 March

1886

- Steamboat *H.M. Carter* (10+ dead), 20 November

1888

- *Steamboat *Sidney* (4 dead, 16 burned), 10 March
- Steamboat *Gold Dust* (17 dead), 7 August

1889

- Steamboat *Corona* ("large loss of life"), Summer

**Burst steam lines*

The steamship J.C. Rawn *was destroyed on December 7, 1939, in Huntington, West Virginia, when two boilers exploded. In addition to serious injuries suffered by the captain and pilot, three crew members were killed.*

1904

- Excursion *General Slocum* (1,021 dead), 15 June

1906

- Steamboat *W.T. Scovell* (10 dead), 20 December

A WORD ABOUT ARTIST ENGRAVINGS

BLOWBACK features numerous engravings as well as rare photographs from the 1800s and early 1900s.

Woodblock engraving was developed by English artist and ornithologist Thomas Bewick at the end of the eighteenth century as a relief printing technique to illustrate birds of Britain. Beginning in the 1820s, engraving evolved into a method of replicating freehand line sketches.

Engravings in periodicals were first employed in 1842 by the world's first illustrated weekly newspaper, The Illustrated London News. *Before publication of engravings, a reader was left to his or her own imagination to envision events only described by a reporter's words.*

The engraving process began with an artist in the field sketching scenes to complement an accompanying news article. An illustrator would then be given the sketch along with detailed instructions from the artist as to how the engraving should be developed. The sketched image was then carefully carved into a wood block used to reproduce the image on paper.

As can be observed, many engravings featured intricate detail suggesting the sketches were quite accurate in their portrayal of the artist's rendering. Today, many early woodblock images are highly valued. Contemporary artisans still use engravings as a medium for their creative outlet.

All of the engravings in BLOWBACK were reproduced from the actual publications in which they appeared. ❂

The Baltimore riot of 1861 took place on April 19 between Confederate sympathizers and members of the Massachusetts militia en route to Washington for Federal service. In a show of force, city police publicly displayed a collection of weaponry — including a steam gun — that could be used against the soldiers. Historians regard this riot as the first bloodshed of the American Civil War.

CHAPTER 17

THE STEAM GUN: Threat or Theory?

I have heard there are troubles of more than one kind.
Some come from ahead and some come from behind.
But I've bought a big bat. I'm all ready you see.
Now My Troubles Are Going To Have Troubles With Me!

– *Theodor "Dr. Seuss" Geisel,*
American writer, 1904–1991

I Had Trouble Getting to Solla Sollew *(1965)*

Black powder, or gunpowder, has been used as a projectile propellant ever since its invention by the Chinese over a thousand years ago.

Although the Chinese used it primarily in rockets (tubes filled with gunpowder), it wasn't until around 1364 the first firearm was introduced. By today's standards, this primitive weapon left much to be desired: Imagine having to light a wick to ignite gunpowder loaded into the gun barrel!

While gunpowder has continuously served as the primary firearm fuel for over 600 years, it is believed there briefly existed in the early 1860s a little-known variation.

Around this time inventors William Joslin and Charles S. Dickinson developed an idea for what could have been defined (back then) as a weapon of mass destruction: a steam-fueled centrifugal-force gun.

When the two Ohio men parted following a disagreement, Dickinson took over the project and relocated to Boston. There he fabricated a rather peculiar, large, cone-shaped iron shield behind which stood a steam-generating boiler and mechanism to fire three-ounce lead balls a range of more than 100 yards.

The working mechanism consisted of a horizontal rotating drum capable of hurling the projectiles through a gun barrel at a rate of 100 to 500 per minute. Activated by steam valves leading from the boiler, the drum was fed balls dumped into a hopper by the operator.

Purpose of the cone-shaped shield was to protect not only the operator from enemy ordnance, but also the launch device and the boiler, serving as propellant source. A horizontal slit extending from the shield's apex permitted lateral movement of the gun barrel during the launch process. Affixed to a four-wheel carriage, the nearly five-ton gun was moved into position by horses.

In 1861, Dickinson publicly and proudly exhibited his steam gun in Baltimore. Alas, his visit to Maryland could not have been more ill-timed.

At this point in history, Maryland found itself literally caught between the Confederacy

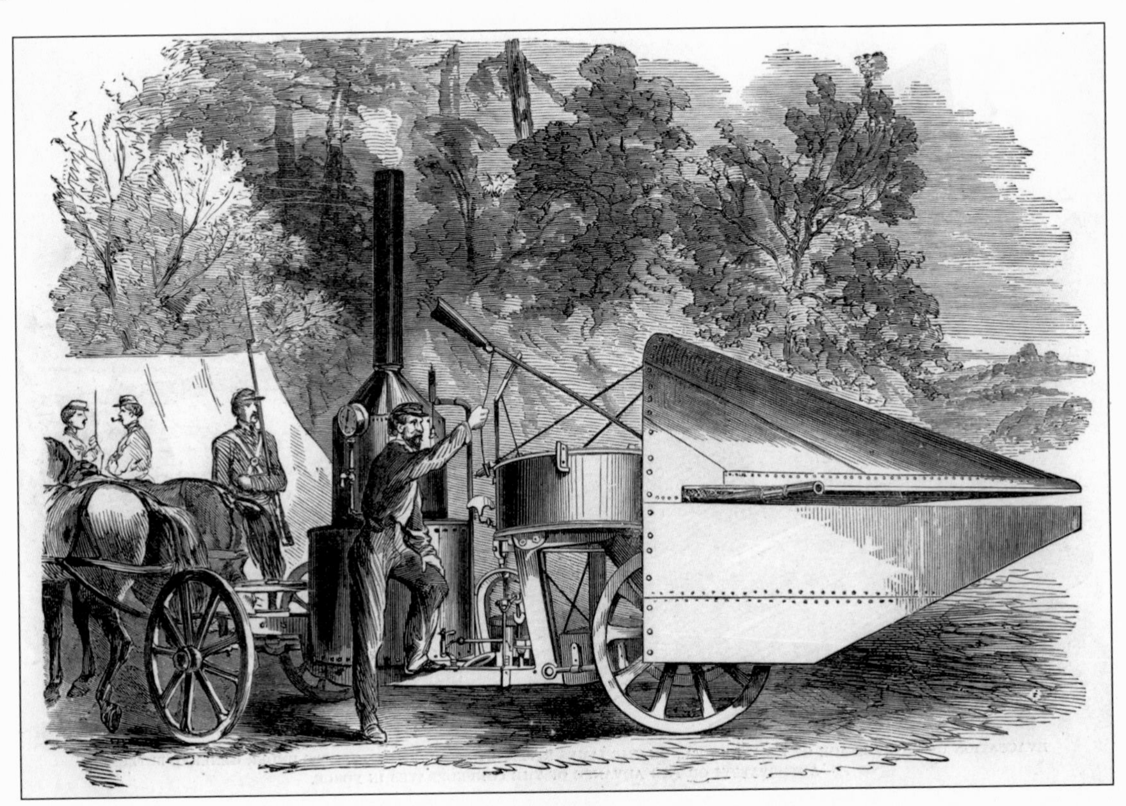

and the North. And while this precipitated a number of problems, Baltimore sympathizers of the South took special umbrage at their police department providing safe rail passage to Union troops from the North through their city en route to Washington. Not lost on the sympathizers was the knowledge that this passage would continue. As tensions flared, everyone knew time was the only deterrent to violence.

States' rights advocates met at Taylor Hall in Baltimore in 1861 to discuss how to deter the sympathizers, thus minimizing anticipated hostilities. At the ready, police decided to assemble all available weaponry. Among the items collected was Dickinson's steam gun.

To communicate their dominance, city authorities publicly displayed a variety of formidable weapons, including Dickinson's gun.

Accounts say the gun was moved to the foundry and machine shop of locomotive and steamship builder Ross Winans, who was charged with the manufacture of pikes (pointed iron spears attached to wood poles), shots, and other munitions.

April 1861 not only marked a significant conflict between Federal troops and secessionists, it launched the steam gun's repute as word of the weapon circulated in newspapers across the country. Strange it was. And it proved to be—allegedly—a powerful deterrent in reckoning with Union forces.

Also making news was 65-year-old Winans, whose wealth, involvement in states' rights issues, and a growing reputation as a manufacturer of munitions made him a national curiosity. His association with the steam gun didn't hurt.

Historians believe this connection was key to the weapon often being referred to as the *Winans steam gun*. (Winans also invented an unusual class of steamships called, most appropriately, *Winans cigar ships*.) Similarly, because of its association with the city, the one-of-a-kind firearm also became known as the Baltimore Steam Gun.

Following the April clash, the steam gun was taken to Winans' shop for repairs before being returned to Dickinson. Perhaps now having a much better idea of the steam gun's worth, Dickinson attempted to move the weapon to Harper's Ferry, West Virginia, where he planned to sell it to Confederate forces. But once again, fate proved unkind.

En route, Union troops captured the gun and moved it to their camp in Relay, Maryland.

There are varying accounts as to how the steam gun finally made its way to Annapolis, Maryland, before being sent to Fortress Monroe, Virginia. The final destination, however, appeared to be Massachusetts, where it was exhibited as a prize of war for a number of years. With the belief the gun was never used beyond the Baltimore riots, it was scrapped sometime before conclusion of the nineteenth century.

As for Ross Winans: His involvement with the steam gun, politics, and munitions manufacture left him with anything but a celebrated reputation. Several days after the Union intercepted Dickinson, Federal forces arrested Winans and briefly detained him for "active sympathy with the rebels." He was released after agreeing to end taking of arms against the government.

While no record exists of the gun being fired, the mere presence of the formidable cone-shaped metal shield was enough to cause some Union opponents to surrender without opposition.

Union forces were unable to determine how the gun worked, supposedly because Dickinson hid vital equipment components.

If able to fire off 300 rounds of two-inch metal balls per minute, as some speculated, Dickinson's gun could arguably be considered forerunner of today's modern machine gun. (The first hand-cranked Gatling gun was patented in 1862.)

While only drawings exist today of Dickinson's invention, a 50-year-old model of the weapon can be found in Elkridge, Maryland, at the intersection of Old Washington Avenue and Route 1.

So much for history.

Did the steam gun work?

It depends on whom you believe. Some said it worked while others discounted the weapon's many laudatory press clippings.

In 2007, the popular cable television series *Mythbusters* constructed a reproduction of Dickinson's armament based on blueprints to test the gun's lethality and effectiveness. The results were less than complimentary:

> The gun performed well on the first two criteria, firing five rounds per second at a range of 700 yards (640 m). However, the weapon lacked any lethal force at ranges beyond point blank, and was not very reliable in terms of delivering the bullets to the targets in an effective manner. The *Mythbusters* concluded that, as a concept and a machine, the steam gun performed perfectly but as a weapon, was too unreliable and impractical.

Even with a colorful and well-publicized history, the steam gun's controversial origins and effectiveness are shrouded in uncertainty.

The only thing certain is: There's no one alive who can corroborate the events of 1861.

CHAPTER 18

PRESSURE EQUIPMENT ACCIDENTS: The Untold and Unusual

(WARNING: GRAPHIC CONTENT)

I have not failed. I've just found 10,000 ways that won't work.
– Thomas Edison, American inventor, 1847–1931

[SOME NAMES HAVE BEEN CHANGED TO PROTECT VICTIM PRIVACY]

Exhaust vents leading from boilers are constructed for one purpose and one purpose only: to provide an outlet for carbon-monoxide-laden air.

But on September 5, 2007, maintenance workers at an elementary school near Anchorage, Alaska, made a grisly discovery while attempting to investigate a strange odor coming from the facility's vent system.

Workers soon discovered the body of 21-year-old Alvin Ricks, lodged at a "T" joint connected to several boilers. Initially speculating a chemical leak when they began disassembling the heating system, workers called police upon discovering Ricks' body.

Anchorage police say Ricks probably climbed into the 36-inch-diameter exhaust from the roof. Once in the vent, he was subjected not only to carbon monoxide but temperatures of up to 300° F.

Sources say the victim had been missing for about two weeks prior to being found. An ID was found in his clothing.

Also found with the body was an alcohol container.

* * *

KESMARC (Kentucky Equine Sports Medicine & Rehabilitation Center) Florida is called "a world class equine rehabilitation facility, dedicated to the recovery and conditioning of equine athletes." Occasionally, part of the rehabilitation process will require an "equine athlete" be subjected to hyperbaric therapy.

Hyperbaric chambers are pressurized oxygen compartments used to decrease healing time of injuries and wounds. These chambers subject patients to higher-than-normal oxygen levels in specially-contained environments.

And so it was on February 10, 2012, event horse Landmark's Legendary Affaire was scheduled for a special session at the rehabilitation center located in Ocala, Florida. The six-year-old thoroughbred gelding from Virginia was being treated for equine protozoal myeloencephalitis, or EPM, a neurological disease.

Initial reports from officials say Sorcha Moneley was in the US to evaluate hyperbaric chambers for possible use in her native Ireland. British-born medical technician Erica Marshall lived at the rehabilitation facility and had operated the recovery vessel for two years.

After the horse was led into the 10' x 12' chamber, Marshall began to increase oxygen pressure inside the enclosed vessel. What was normally an uneventful procedure became unpredictable when the horse began kicking. As the steel-shoed animal robustly pummeled the chamber's inner wall, Marshall reacted by decreasing the internal pressure. But the damage had already been done.

The kicking chipped the chamber wall's protective coating revealing a metal surface. As one of the steel shoes subsequently struck the exposed metal, a spark ensued, causing a pair of explosions that launched debris over a 1,200-square-foot area.

It was believed Marshall and the gelding were both killed instantly. Moneley, who just left Marshall to obtain help with the kicking thoroughbred, was transported to a health care center with traumatic injuries.

* * *

In July 2007, a 36-year-old man was killed when lightning struck his air tank as he surfaced from scuba diving near Deerfield Beach, Florida.

Police say Josh Lang was a first-time diver who was accompanied by three friends.

According to authorities, Lang had surfaced about 30 feet from the boat when the tank was struck.

An autopsy revealed the victim died instantly of electrocution.

In May 2010, a woman from Xiamen in China's Fujian province was critically injured but survived when her chair exploded as she worked on her computer. Doctors removed a dozen screws and fragments of plastic from her flesh.

* * *

A 14-year-old boy from Jiaozhou, China, was killed in 2009 when the gas cylinder chair on which he sat exploded.

Reports say the blast propelled sharp pieces of the chair into the victim's rectum, thus causing extensive bleeding.

The highly pressurized gas cylinder chair is not unlike those found in many offices where the cylinder raises and lowers chair height.

Alone at the time, the 170-pound victim was able to summon an ambulance but apparently arrived too late at the hospital, where he was pronounced dead.

The hospital reported three similar exploding gas cylinder incidents within a recent one-month period, thus indicating a number of poor-quality chairs being used in the area.

A similar accident occurred in 2007 when a 68-year-old man survived being impaled by a nearly eight-inch-long chair part that created a two-inch wound.

It is believed any of three factors could be responsible for the rash of office chair explosions: gases other than nitrogen being mixed within the cylinder, substandard materials used in cylinder construction, and lack of an airtight seal.

Oil-based hydraulic cylinders are reported to be safer than pressurized gas. However, many office chairs today employ gas cylinders. Most of the latter are constructed in China, where each of the aforementioned incidents occurred.

* * *

Two burglars in Britain received a rude greeting while attempting to steal a boiler in October 2006.

Following a massive explosion in West Bromwich, police found one man dazed under a pile of rubble and another in the garden of what used to be a house. Both were seriously injured.

West Midland Police Superintendent Bob Spencer said the burglars may have been cutting through a gas pipe to remove the boiler. "One [burglar] was found in the back garden under the rubble, so it must have blasted him from the first level of the house." A fire department spokesman stated it took rescuers an hour to extricate the other suspect trapped within the house.

According to a bystander who heard the explosion, "The house had been boarded up for ages and people go in and steal the copper piping. There was a big bang and then a vibration that I could feel in my car."

As a precaution, hundreds of homes within relative proximity of the destroyed house were evacuated. Both suspects were taken to area hospitals and treated for serious burns.

As to how the explosion occurred, Superintendent Spencer speculated: "Who knows, they just might have been smoking."

* * *

On the morning of St. Patrick's Day 1909, a group of people waited for loved ones in the ladies' waiting room at Montreal's Windsor train station.

Like a half-million Boston arrivals and departures earlier, the public mood in the waiting area was rife with anticipation. Constructed in 1889, the passenger terminal was not only a Montreal landmark, but a proud symbol of the city's heritage.

As locomotive 2102 made the final approach to the station, an explosion rocked the engine car, causing a lurching effect that forced a driving wheel into the already damaged boiler. The escaping steam and hot water severely scalded engineer Mark Cunningham and fireman Louis Craig, who both jumped from the train in an act of self-preservation.

Unbeknownst to the 200 passengers in cars behind the engine, the train lost no speed barreling without an operator to its final destination. When a conductor signaled the engineer a passenger was to be let off a station before Windsor, he became concerned when no response was received.

Picking up speed estimated to be between 50 and 55 mph, the runaway train crossed several Montreal blocks before a brakeman activated the emergency air brakes. But the brakes failed to prevent the engine and its cars from smashing through the safety buffer at track's end. Traveling at about 25 mph, the train continued into the ladies' waiting room before coming to rest in the station's main concourse, the front of the engine breaching the depot's southernmost wall.

While there were no fatalities on the train, passenger W.J. Nixon discovered his wife and two children perished when the engine slogged through the waiting area. Twelve-year-old Elsie Villiers was also killed instantly by building debris collapsing all around her from the horrendous impact. Engineer Cunningham died from his injuries a few hours after the incident at a Montreal hospital. Eleven others were critically injured.

The accident inspired song writers Henri Miro and Raoul Collet to write the musical piece *La catastrophe de la gare Windsor* (The Windsor Station Disaster).

* * *

In March of 1899, factory workers in Pittsburgh took leave of their jobs at the noon lunch hour. What followed within moments was irony of significant consequence.

A boiler explosion killed five men, wounded 12 others, and reduced the large one-story building to a pile of shattered brick and mortar. Each of the victims was inside the factory when the accident took place.

As newspapers would later report:

> The boilers were inspected six months ago, and were thought to be in first class condition. The real cause of the explosion will probably never be known, as the

> engineer was killed outright, and no one has been found who was in the engine room at the time the disaster took place.

While such scenarios were rather common around the turn of the century, the location was not.

Among those who died at West Point Boiler Works were four boilermakers.

* * *

In January 1912, Joseph Albrecht decided to visit the railroad heating plant where he had worked for several years.

Because of advancing age, the 67-year-old former laborer had been replaced a few days earlier by 29-year-old immigrant Fred Streck.

Streck and his predecessor were engaged in conversation when a high-pressure pipe ruptured and scalded both men with steam. Although Albrecht was burned about the face, head, hands, and arms, it was the younger Streck who took the brunt of the accident.

Concerned for Albrecht's age, railroad authorities nonetheless agreed the veteran worker was physically better off than his successor. A somewhat dazed Albrecht managed to walk several blocks back to his home. Subsequently washing his face and hands, Albrecht shared details of the accident with his family as well as his concern for the fate of Streck (who unbeknownst to the elder laborer had already died).

Shortly thereafter, the older man slipped into a state of semi-consciousness and expired that evening.

The irony of the former and present holder of the same job perishing as a result of the same incident is not lost. For Albrecht, it was simply having been at the wrong place at the wrong time.

Yet Albrecht's cause of death was not attributed to the large amount of steam he had reportedly inhaled. It was shock complicated by old age.

* * *

While most pressure equipment carries significant potential for explosion, danger is not necessarily limited to internal workings.

Ask Jonathan Metz.

Upon performing some routine maintenance in June 2010, the West Hartford, Connecticut, resident found himself in the awkward position of being unable to extricate his left arm from the funnel-shaped fins of his basement boiler. Awkward in the sense he could not sit on the floor or completely stand upright. For two days.

About 24 hours after he became stuck, Metz accepted the reality he was going to die: in his basement, with left arm—shoulder to elbow—one with the boiler. And discovered by who knows whom.

As he listened to the restless stirrings of his dog Portia upstairs (also without food and water) Metz appealed to a higher power.

A window opposite the boiler was unreachable, as were most of the tools in his modest basement workshop. Isolated in a way that can only be imagined in a bad horror movie, the Connecticut man suddenly became inspired by sipping a few musty droppings of boiler water.

"It gave me enough mentally to believe that I could go on," he recalled. It gave me a boost, that here's water—it's enough to at least get by."

With enough time to reflect on his destiny, Metz thought about the blessings that make life special. Such as his fiancée. His parents in North Carolina. And of course, the now seriously neglected Portia. How would *they* view his unfortunate demise?

Having tried every movement humanly possible to free the trapped appendage, tearing of flesh precipitated the development of gangrene. It became apparent if he was to do something, it needed to be drastic. And soon.

Able to reach several saw blades, the 31-year old quickly determined there was but one option: self-amputation. But identifying difficult choices and summoning one's courage are seldom easily reconciled.

Metz converted his shirt to a tourniquet, pulled it as tightly as his strength would allow, and commenced surgery.

"I would say about 90 percent of the cut was surprisingly pain-free. The pain was in having to look at it and see it and see what you were doing, or what I was doing to myself. But physically, it wasn't that bad."

That is, until he reached the nerves on the arm's underside. "I can't even describe the pain—nothing I've ever felt in my entire life—I tried cutting a little bit more, hit another nerve, and again—lightning bolt-like pain. And that's when I just said: I can't do it. I can't finish this cut."

His feeling of death now certain, Metz started to scrawl a note on the boiler to his fiancée and family—in his own blood.

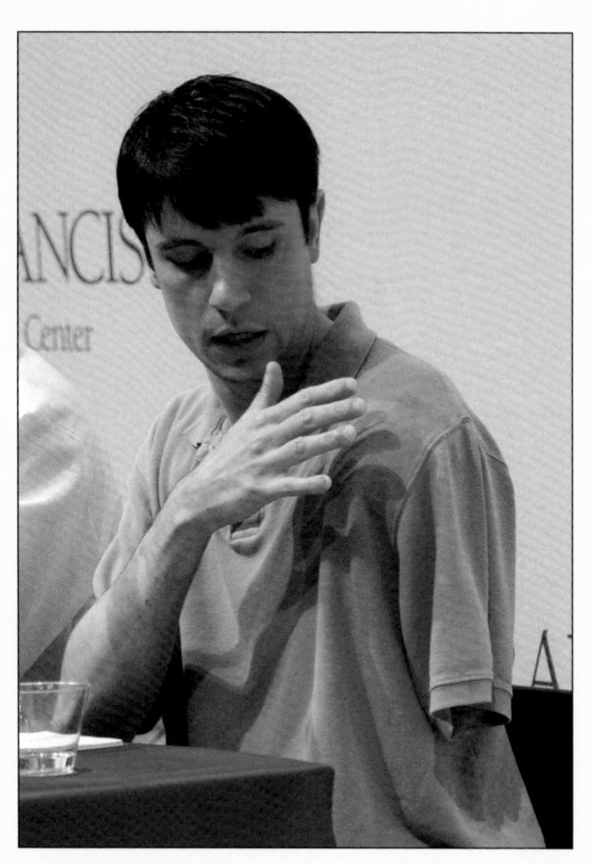

Jonathan Metz.

After his second day in the basement, Jonathan Metz was rescued by a friend who became concerned when Metz did not answer the door.

While he never completed the amputation, doctors say Metz avoided blood poisoning by slicing away the diseased flesh. Doing nothing would surely have killed him.

While the boiler has been replaced, blood on his basement floor remains—a visible reminder of the power of perseverance. And prayer.

* * *

Few realize just about everyone in the civilized world comes into close proximity of a pressure vessel several times each day.

But some British coffee drinkers in the town of Farnborough are now better informed than most.

Shortly after noon on September 14, 2010, patrons of the community's Sainsbury's supermarket café were dazed by a deafening blast caused by an espresso machine boiler. Six people were taken to an area hospital with minor injuries.

According to eyewitnesses, the explosion blew a hole in the café wall and destroyed the coffee counter. Patrons were showered with cups and saucers. One witness explained, "There was a loud bang and the machine shot 20 feet off the counter. The boiler had burst and sent hot water and steam all over the place."

A young woman was treated for head, arm, neck, and eye injuries. Other victims reported cuts and burns.

Officials evacuated the store as a safety precaution only to find the weighty 4'x 3' machine a mass of mangled tubing. Environmental health officials were called in to investigate.

A supermarket spokesperson speculated the boiler was blown from its mountings as a result of "over-pressurization." More specifically, she said, it was caused by the failure of a pressure relief valve.

The espresso machine's manufacturer attributed the problem to a design flaw.

* * *

Warnings seem to be posted everywhere these days. As bothersome and, at times, mundane as these might seem, many caution signs are intended for people who don't pay attention to caution signs.

Consider the Michigan man who—while hauling filled gas cans in the back of his minivan—experienced a sudden urge to smoke. One click of his lighter was all it took to ignite fumes that accumulated inside the passenger compartment.

As one of the cans exploded, both driver and passenger were set afire. Escaping the burning van both men made a purposeful dash toward a lawn sprinkler. The now unmanned minivan continued its journey through a wooden fence, and across the driveway of a fire station, where it came to a stop before striking the stationhouse.

Unaccustomed to fires coming to them, firefighters from the station worked to extinguish the burning fence as well as the fully consumed van.

Officials said both occupants of the van were expected to make painful recoveries. They did not release their names.

* * *

Pressure cookers have been used for decades to steam vegetables, meats, and potatoes to tender and tasty perfection.

On May 19, 2011, at around 10 A.M., 79-year-old Mirta Debesa readied her pressure cooker to steam some red beans. Having used the cooker numerous times, Mrs. Debesa knew her way around the seemingly benign appliance.

That all came to an end when the Miami woman's son heard a loud explosion. Rushing to locate the blast, Martin Debesa entered a bean-splattered kitchen with parts of the cooker scattered in several directions. On the floor lay his mother in a puddle of blood, her foot severed above the ankle.

Mr. Debesa reacted by tying a belt around his mother's damaged limb in an attempt to stem the bleeding. As the son later recounted: "It was like a bomb. My mother was lying there; her left foot was severed, cut more or less to the ankle. There was blood rushing in. I got panicky for a little while, but I recovered myself."

Rescuers took Mrs. Debesa to a local hospital where she was listed as stable.

Back at the Debesa home, investigators had more questions than answers.

"Right now, we don't really know exactly what happened," opined Lt. Ignatius Carroll of Miami Fire Rescue, "but from what we are told, she was cooking, and for some unknown reason, the pressure cooker fell. The impact caused it to explode."

Carroll termed the morning incident as "one of the freakiest accident calls that we have ever heard."

Officials agreed: if the Miami woman had been alone, and without the application of a tourniquet, she could have bled to death. The portion of the cooker that struck Mrs. Debesa was not identified.

A week after the accident, the 79-year-old mother was transferred to a nursing home. Two weeks later, less than a month after the fateful incident, she died.

Cause of death was not revealed.

* * *

In March 2010, a married couple in Britain prepared to leave for the evening when an explosion occurred in their home, shattering windows as well as lifting the roof from the structure's frame.

According to the watch manager at the fire station, "The whole roof came up like a lunch box lid."

The detonation was centered in a home extension where a leaking aerosol can created a flammable atmosphere, and thusly a small fire.

"Once the boiler kicked in, it sparked the explosion," reported the watch manager.

In what can only be called a "freak accident," firemen cited luck as the reason no one was injured.

* * *

In the nineteenth century, boiler explosions were to riverboats what tornadoes are today to mobile homes. And so it was in 1872.

Following months of waiting for the Darling River to rise, the paddle-steamer *Providence* finally achieved flotation and headed south with 200 bales of wool. Approaching Kinchega homestead in New South Wales, Australia, the steamer's boiler exploded, killing four crew members.

Force of the concussion not only split the boat's hull, it launched an anvil and heavy hammer yards from the disaster scene.

Legend has it that before their fatal departure, the crew gathered at Menindee pub before returning to the paddler and igniting the boiler for their long-awaited departure.

But was a failure to check the boiler's low water level a result of the crew members' intoxication?

Not according to Bob Butrims.

After examining the boiler in question in 1996, he concluded it was of faulty construction, which meant the *Providence* "was a time bomb waiting to go off."

So how was Butrims able to locate a historic yet damaged piece of pressure equipment more than over 100 years later?

It can be found today where it landed: embedded along the Darling riverbank several hundred yards from where the explosion occurred.

* * *

An ear-shattering blast at Montour Rolling Mills on Thursday, October 8, 1896, rocked the small Pennsylvania community of Danville.

It was just before eight P.M. when the No. 5 boiler exploded, catapulted through the factory, and left the plant via a newly ventilated wall that failed to impede or divert the equipment's fateful journey. The 28-foot boiler continued on its route 175 feet until crashing through a house and finally coming to rest 100 feet beyond.

Following the ensuing confusion, it was learned the boiler killed a 6-week-old baby being held in the arms of his mother, Mrs. John Baron. Mother and baby were in the house as the large metal projectile entered and exited the home on Northumberland Street. Mrs. Baron suffered several broken ribs.

Back at the plant, five workmen died as a result of the explosion. Another 33 employees, including Mrs. Baron's husband, suffered a wide range of injuries. Witnesses said the explosion also sent part of the boiler toward the plant's interior, where workers ran to elude scalding steam, brick chunks, and shards of hot metal.

The plant experienced considerable damage but reopened just two weeks following the tragic turn of events. The house was completely destroyed.

Cause of the accident was never determined.

Strangely, the October 8, 1896, incident occurred almost 42 years to the day following an earlier catastrophe at the same mill. On October 7, 1854, a lethal boiler explosion killed 10 people and injured scores of others. That boiler was also blown from the plant. It, too, struck a house—right next door to the home that would be destroyed by the airborne boiler in 1896.

The 1854 disaster killed two children asleep in their beds. One of the small bodies was discovered inside the boiler.

* * *

In March 2011, Janice Broaden loaded her small sport utility vehicle (SUV) with inflated balloons destined for a baby shower. Stopping at a store en route, the Union Springs,

Alabama, woman returned to her car and opened the passenger door.

As she did so, the interior of the vehicle exploded with such force it instantly denuded the woman before violently thrusting her to the ground.

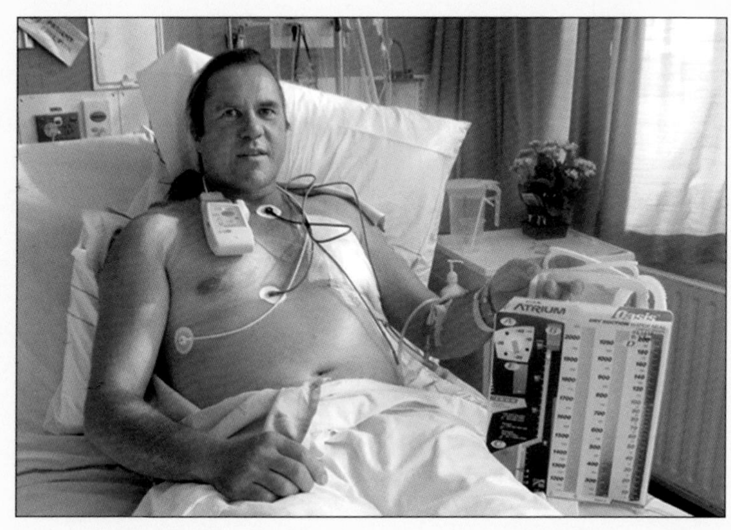

Steven McCormack.

According to authorities, the blast could be heard as far away as four blocks. Some felt the ground shaking two blocks away. Initial reports indicated a helium tank exploded.

Ms. Broaden was transported to a Birmingham hospital, where she was treated for second-degree burns and placed in intensive care. A second victim, Elaine Taylor, stood only a few feet from the blast and was treated for minor injuries.

With blown-out windows and a bulged passenger compartment, the SUV was totaled.

When a helium tank could not be located, investigators focused on the gaseous content of the balloons. Investigator Louis Murry explained the inflatables were filled, not with helium, but pure oxygen.

The Alabama Fire Marshall's office concluded the blast resulted from the balloons being overheated.

NOTE: Theories regarding what type of gas could cause the Union Springs explosion are not without controversy. Some experts claim the violent reaction must have involved hydrogen or perhaps a mixture with oxygen. One thing is certain: Helium would not have precipitated such an explosion. The presence of oxygen is necessary for the ignition of flammable gases.

* * *

Now consider the dilemma of Steven McCormack, a truck driver on the east coast of New Zealand's North Island.

Standing on his vehicle's running board in May 2011, the 48-year old suddenly slipped and fell onto the brass fitting of an air hose leading from the truck's breaking system air reservoir. As he landed on his back, the broken compressed air hose punctured his buttock, instantly introducing air into his body at 100 pounds per square inch.

The stunned McCormack admitted his only option was to lie on the ground.

"I felt the air rush into my body, and I felt like it was going to explode from my foot," he later recalled "I was blowing up like a balloon."

Hearing McCormack's screams, co-workers ran to his aid, one of them releasing the safety valve to halt air flow.

The New Zealander was whisked to a hospital for treatment of body swelling and fluid in his lung. Doctors said McCormack was fortunate the air did not enter his bloodstream, although it did separate the fat from his body's muscle.

* * *

Working as a pilot on the steamboat *Pennsylvania*, Sam Clemens did not get along with associate pilot Bill Brown.

Deciding to put the unpleasant relationship and the *Pennsylvania* behind him, Clemens departed on June 5, 1858. But before leaving, he arranged a job for his younger brother, Henry, as the vessel's "mud clerk."

Eight days later as the *Pennsylvania* steamed near Ship Island south of Memphis, and while the ship engineer was alleged to be preoccupied by lady friends, a boiler explosion ripped through the vessel with stunning rapidity.

Henry was severely scalded about the skin and lungs. He succumbed on June 21.

But before his brother's death, Sam managed to spend time with Henry. He would later explain: "For forty-eight hours I labored at the bedside of my poor burned and bruised but uncomplaining brother and then the star of my hope went out and left me in the gloom of despair."

It was said Sam was never the same after his brother's passing. Years later, nor would his name: Mark Twain.

* * *

There are reasons they are called *flammable* gases.

With proper usage, hydrogen, propane, methane, acetylene, ethane, and ethylene are effective energy sources in an industrial environment. Under the wrong conditions, however, these volatile substances can ruin what otherwise might be a perfectly fine day.

An excited Alan Nelson of Colorado and his girlfriend thought it would be novel to celebrate Super Bowl XL in 2006 with an impromptu pyrotechnic display at a friend's party. And what better way, they thought, than filling a balloon with acetylene, a flammable gas often used in welding.

In what can only be categorized as balloon buffoonery, police theorized that while the inflatable vessel was being transported, it rolled across the back car seat creating static electricity and thus igniting the gas.

Shortly after the incident, Arapahoe County sheriff's deputies spotted Nelson's car abandoned behind an old gas station. In addition to windows being blown out, the vehicle doors were bulged outward. The blast also added about a foot to the car's headroom when the roof was pushed to its structural limit.

Deputies traced the car license plate to Nelson, who confirmed the explosion and the kindness of a passer-by who provided the couple a ride home. An ambulance subsequently took the 46-year-old and his girlfriend to an area hospital, where they were treated for shrapnel wounds and ruptured eardrums.

* * *

While the game of paintball involves figuratively eliminating one's opponent by splattering him with paint-filled gelatin ammunition, the possibility of actual death seems infinitesimally remote.

However, CO_2 canisters (i.e., small, high-pressure aluminum tanks) used in paintball guns create enough force to propel the 14-mm paintballs 250 to 300 feet per second, or at speeds up to 200 mph.

On June 7, 2003, Brandon Johnson couldn't wait to use his new paintball gun for the first time. Although he only had it for three days, the 15-year-old and five friends took to the back woods in Olympia, Washington and put on their protective gear.

Two hours passed without incident.

Returning to a friend's front porch, Brandon and his paint-stained buddies started to remove their CO_2 canisters. Unfortunately, Brandon had also removed his helmet and eye protection.

As the Washington teenager unscrewed the cylinder, it separated from the valve and struck him in the forehead. The high-velocity projectile continued in an unknown direction and was never located.

Brandon died five days later from injuries he sustained as a result of the violent tragedy.

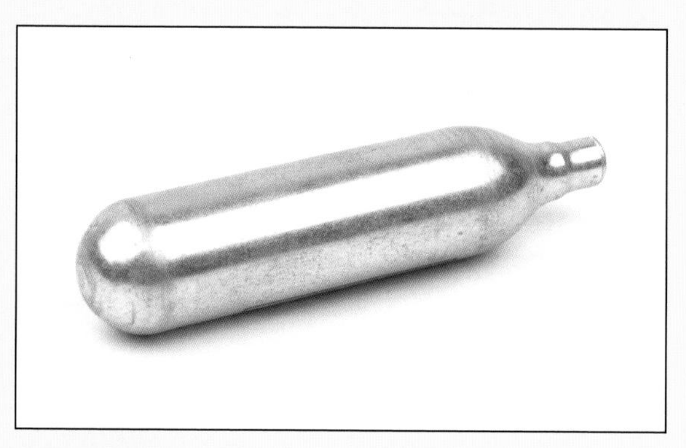

CO_2 capsule.

In a similar circumstance, Colette Contois was delighting in a get-together of her young son's friends on February 16, 2004.

And what a get-together it was. Taking place at a paintball field near Swansboro, California, it was a long-awaited event for all who attended.

At day's conclusion, the young participants started to disassemble their guns in an area run by the paintball facility.

As Colette and her husband watched over the activities, one of the young guests detached a cylinder valve from the cylinder (not the gun, as intended), instantly converting the aluminum canister into a deadly missile. The errant projectile struck the California woman in the back of the head with stunning rapidity. She died that day from her injuries.

Occasion of the gathering: Her son's 10th birthday.

* * *

It is long held the commission of suicide with a deck of playing cards is an urban legend.

In reality, however, such an unlikely occurrence was indeed factual, and as bizarre as it was disturbing.

San Quentin inmate William Kogut, condemned to death row for slashing the throat of rooming house operator Mayme Guthrie, was awaiting execution on October 20, 1930.

Possessing a pack of playing cards, Kogut reduced the deck to small bits by tearing each card by hand. He stuffed the pieces tightly into a hollow steel leg removed from his cot, and plugged one end of the tube with a broom handle. Kogut soaked the ripped cards by pouring water into the open end.

The convict then placed the device on a kerosene heater next to his cot, with the open end of the tube against his head. The process of heating and then converting the water to steam generated enough pressure to eject the card fragments with sufficient velocity to penetrate Kogut's skull, causing instant death.

According to authorities, Kogut felt compelled to end his own life rather than affording that pleasure to the state.

* * *

Actress Sirkka Sari played the lead role in the 1939 film *Rikas tyttö* (rich girl).

Following a day of shooting at the Aulanko Hotel in Hameenlinna, Finland, and with the production nearly complete, the Finnish Thespian joined a lively party of cast and crew.

Sirkka Sari.

Sometime during the festivities, and for unknown reasons, Sari accompanied a man to the hotel's dark, flat roof. There she spotted a chimney with a ladder leading several feet to the top.

Thinking the chimney was a theatrical prop, the 19-year-old climbed the ladder and fell into a heating boiler where she died instantly.

Rikas tyttö was her third film. And unfortunately, her last.

* * *

James LeBeau was described by his mother as "a kid who tinkers with things ever since he was little." At 24 years of age, the Kent, Washington, man was still tinkering.

And so it was in November 2004 that LeBeau began toying with a lava lamp at his trailer home.

Heated by a small light bulb, a lava lamp is generally innocuous. Filled with a mixture of wax or oil and water, enclosed in a glass bottle, the iconic lamp of the 1960s and 1970s also contains a small pocket of air to allow the liquid to expand.

Unable to contact James at his home, parents Herm and Janice LeBeau drove from their residence in Auburn, Washington, to check on the youngest of their three sons. What they discovered left them in shock.

Quickly surveying the home, Herm LeBeau discovered his son's body slumped in a corner near his stove.

"There was glass from the kitchen clear to the living room," explained Janice LeBeau.

The grieving mother theorized her son placed the lava lamp on his kitchen stove. "It wasn't bubbling fast enough for him," she surmises.

Authorities believe additional heat from the stove increased pressure within the lamp, triggering a detonation not unlike a grenade. A large shard from the shattering glass was driven deep into the young man's chest.

It was further revealed James LeBeau never had an opportunity to call for help and bled to death.

Police found no evidence of drug or alcohol use. The King County medical examiner ruled the death accidental.

* * *

At four A.M. on November 5, 1983, four divers occupied a decompression chamber system attached by a short passageway to a diving bell on a semisubmersible drilling rig in the Norwegian sector of the North Sea.

As one of the divers was preparing to close the door between the system and passageway, the chamber was explosively decompressed from a pressure of nine atmospheres to one—all within a split second.

The body of the diver exposed to the highest pressure gradient violently exploded from the rapid and massive expansion of internal gases. His remains, including his thoracic and abdominal organs, as well as his thoracic spine and limbs, were forcefully expelled through the 24-inch diameter opening left by the jammed chamber door.

Portions of the body were discovered all about the rig with one part found on the rig's derrick, 10 meters directly above the chamber.

Authorities speculated death was painless and instantaneous

Counting two dive tenders, five men were killed, with another severely injured.

* * *

By all outward indications, Tracy Kraling appeared to enjoy her job at the Regions Hospital animal research laboratory in St. Paul, Minnesota. Among her duties: cleaning cages and other equipment in a walk-in sanitizer, which utilized steam to sterilize instruments to 180°F.

Constructed in 1980, the autoclave operated similarly to a dishwasher. Closing the door activated the "wash" cycle.

On November 4, 2004, while working alone, Kraling became trapped in the autoclave when the door closed behind her, triggering the automatic cleaning process. She was unable to deactivate the large machine or extricate herself, all the while being exposed to jets of saturated steam.

When discovered by colleagues, the 31-year-old Roseville, Minnesota, woman had already suffered severe scalding. She succumbed one day later as a result of her injuries.

Minnesota's Occupational Safety and Health Administration fined the hospital $75,000 because the autoclave had no escape route, panic bar, or accessible off switch. The agency emphasized the sanitizer should not have been able to activate while entrapping the worker within.

At the time of the accident, Tracy Kraling had been married six weeks.

* * *

Among those killed in December 1850 by a boiler explosion onboard the side-wheel riverboat *Knoxville* was inventor Alfred Stillman. The New Yorker held hundreds of patents, including one for extracting juice from sugar cane.

As the *Knoxville* was leaving the wharf at the foot of Gravier Street in New Orleans, a blast of spectacular fashion rocked the city for blocks.

The boiler explosion collapsed both flues before destroying the boat's upper works forward of the wheelhouses. One of the boilers was launched through the cabin of the nearby steamer *Martha Washington* and onward into the ladies' cabin of the steamer *Griffin Greatman*. The other was catapulted 150 yards across the levee, knocking down large piles of flour barrels before passing over the heads of a crowd gathered on the wharf.

Casualties included 18 passengers onboard the *Knoxville*, most of whom were either killed or missing.

Of inventor Stillman's 572 patents, ironically one was for a steam boiler safety fuse. Invented in 1847, the fuse was simply called "an improvement in preventing Explosions of Steam Boilers." ⚙

CHAPTER 19

HUMAN CANNONBALLS: High-Caliber Performers

A bend in the road is not
the end of the road...unless
you fail to make the turn.
– *Helen Keller,*
American author/activist/
educator, 1880–1968

And so, it may be asked, what does shooting a person out of a cannon have to do with pressure equipment?

First, let's dispel any perception human cannonballs are propelled by gunpowder.

As a matter of fact, there never has been anyone who has been launched by cannon fire. Or at least anyone who has lived to tell about it.

Comparatively speaking, being shot from a cannon is the easy part. Landing in a net, well, that's where it becomes rather dicey.

Historian A.H. Coxe reveals that over the years, more than 30 human cannonballs have met their demise. Many succumbed by failing to reach or overshooting the intended destination: a target net or airbag. While 30

Lady human cannonball Egle Zacchini, 36, attired in siren suit, crash helmet, and asbestos mask prior to launch in 1948. Photo by Cornell Capa.

fatalities may not seem prodigious, there have only been approximately 50 daredevils who have braved this circus event.

Brave, however, might be a bit of an overstatement. Circus lore is rife with tales of cannonball calamity. But make no mistake: This is not a profession for dummies.

In addition to having limber, aerodynamic bodies, these human projectiles must be able to calculate with precision where they will alight. Traveling unsuspended as high as 100 feet at a speed of 60 (some say as high as 90) mph allows for nothing less than exact precision.

Among the considerations: the cannon's position and angle, and the daredevil's parabolic trajectory (initial velocity at a radians angle from the horizontal).

According to the Ontario Science Centre, other factors include:

Weight Of The Performer

Different performers have different body weights, which affect the calculations. Also, a performer's weight can fluctuate from one performance to the next.

Air Resistance

To decrease air drag, performers must wear tight clothing.

Wind Speed

Winds can slow down or accelerate a projectile, or even push it sideways.

Aerodynamics

Tumbling in flight can shift a performer's center of gravity, altering the trajectory and landing.

So what precisely happens when the cannon is fired and discharges its human payload?

No one knows. Exactly.

Those in the profession going back to the 1870s have never shared how it's done. Such is protecting one's secrets from the competition.

But enough has been observed over the years, and this is where pressure equipment is particularly relevant.

Time lapse photo of human cannonball en route to net at the Royal Adlaide Show.

HUMAN CANNONBALL ABBREVIATED TIMELINE

1871

"The Great Farini" (William Leonard Hunt) patents a device for launching "human projectiles." Based on a system of rubber springs, it is not enclosed in a cannon barrel.

1873

A young man posing as a woman named "Lulu" is the first to try Farini's invention. The act is called the "Lulu Leap" and features the performer jumping nearly 40 ft. (12 m) into the air.

1877

Farini, as the entertainment manager of the Royal Westminster Aquarium, introduces human cannonball "Zazel."

1879

The legendary showman William Leonard Hunt patents the cannon design.

1882

Farini and Zazel appear in P.T. Barnum's "Greatest Show on Earth."

1890s

The public loses interest in "human cannonball" performances. Loop-the-loop bicycle acts become popular.

1920s

Paul Leinert and several other European performers and inventors further develop and improve the cannon design.

1929

John Ringling sees the Zacchini family perform their human cannonball act in Europe and brings them to America. The father, Ildebrando Zacchini, creates a new cannon powered by compressed air, accompanied by smoke and a loud charge.

1932

Mario and Emanuel Zacchini (two of Ildebrando's seven sons) invent a double cannon that fires two people in rapid succession.

1939

The Zacchinis leave Ringling Circus and open their own traveling carnival. Their new act sometimes features both men being fired over the top of Ferris wheels.

1940

Mario retires after sustaining severe injuries during a performance at the World's Fair in New York.

1995

David Smith Sr. breaks the distance record set by the Zacchinis. Smith is shot 180 ft., 4 in. (54.97 m.) at a performance in Manville, New Jersey.

1998

David Smith Sr. and David Smith Jr. perform in a "Duelling Cannons" event for the Guinness Book of World Records. Smith Sr. breaks his own earlier record with a shot of 56.64 m (185 ft., 10 in.), while his son successfully completes a shot of 55.19 m. (181 ft., 1 in.).

Simply explained: The human cannonball climbs into the cannon and positions himself on a platform located at the bottom of the big gun's barrel. Situated just above an air pressure tank, the platform moves freely on a rail within the barrel and is the actual vehicle launching the cannonball performer.

At show time, smoke and sound effects simulating a real cannon blast are initiated. At 150–200 pounds per square inch, the compressed air instantly drives the circular platform to the top of the barrel. And while the cylinder stops at the barrel's end, the passenger continues a journey that ideally ends on the netting or airbag.

Naturally, the possibility of something going awry is always present. A typical human cannonball must continually train while maintaining strong legs, knees, and back to withstand punishing launches.

Cleveland State Professor of Physics Jearl Walker estimates air pressure accelerating the launch propels the human cannonball at nine times the force of gravity. This force instantly moves blood from the brain to the feet and knocks most people unconscious. It

A human cannonball emerges from a cannon at Boston's Topsfield Fair, circa 1930s.

also compresses the backbone to the degree that performers temporarily lose two inches of height.

"The trick is to regain consciousness before they [performers] hit the nets," Walker explains.

Until 2011, the distance record for a human cannonball was held by David "Cannonball" Smith Sr., who accomplished the feat in 1998 by traveling 185 feet 10 inches (56.64 m).

A former junior high school math teacher, Smith says of his cannon, "There's enough power in there to make peanut butter out of you." As for the net, he highly recommends landing on one's back: "If you were to land standing up, you would have hips for earmuffs."

An air pressure tank and cannon of precise construction do not preclude wear and tear on parts or mechanical breakdown. In 1999, cannonball performer Brian Miser arranged overseas shipment of his cannon for a show in Japan. There he missed his air bag and smashed into a scaffold. Having broken his pelvis, three ribs, and a foot, Miser later discovered a tiny crack in the cannon, which caused a small change in his trajectory.

While a human cannonball's job may not be for everyone, it is a profession with roots going back to the 1870s.

Among the first human cannonballs was a 14-year-old girl who went by the name "Zazel" (Rosa Matilda Richter). In 1877, she was launched by a spring-style cannon developed by George Farini (William Leonard Hunt).

Perhaps the most recognized name associated with human cannonball acts is the Zacchini family. In the 1920s, Italian acrobat Ildebrando Zacchini was credited with refining Farini's invention by introducing an air-powered cannon.

It appears, however, it was one of Ildebrando's seven sons who in fact created a new propulsion source for the exceptionally popular Zacchini family act.

During his service in World War I, Edmundo Zacchini gave considerable thought to the infiltration of enemy soldiers. One idea he seriously pondered was shooting troops from cannons across enemy lines.

Although his idea was never seriously entertained, Edmundo set out to test his

Fourteen-year-old acrobat "Zazel" (Rosa Matilda Richter) before her performance in 1877 at the Royal Westminster Aquarium, London

concept following the war. Employing two 23-foot barrels, he carefully placed them inside a smaller-diameter barrel. Using compressed air as his energy source, Edmundo frequently tested the one-ton cannon. He broke his leg five times, thus leaving his right leg four inches shorter than the left.

(Edmundo's scheme to use cannons for human transportation may not have been as bizarre as it was perceived during the war. The concept of accelerating delivery of special forces, police officers, and fire fighters onto the roofs of tall buildings by way of an 80° angled ramp with side rails has been submitted for patent consideration by the U.S. Defense Department's Defense Advanced Research Projects Agency, or DARPA. Seen as a variation of the human cannonball cannon without a long barrel, the computer-controlled, four-meter-tall ground-level launcher could supposedly transport a man to the roof of a five-story building in less than two seconds.)

Cannonball performers say the public still embraces their death-defying four-second flights. And that may be why they still exist despite a woeful mortality rate.

Twenty-eight-year-old Brazilian stuntman Diego Zeman grew up around circus performers. For years, his Brazilian father and Hungarian mother performed as one of the top acrobatic acts in Eastern Europe.

Having worked out and performed as an acrobat since age eight, Diego got into the human artillery profession when the human cannonball before him was suddenly dismissed.

Performing as a cannonball artist, Diego explains, is done "for the enjoyment, not the money." That "enjoyment" requires the stuntman to spend quite a bit of time exercising and maintaining his weight.

To prepare his body for what can only be described as a physically grueling, high-speed free fall, Diego attended a course at the Guyana Space Center in French Guyana.

Diego admits the South America trip was really responsible for him getting into the profession—thanks to the sudden dispatch of the stunt performer before him.

His predecessor, Diego notes, wanted to attend the space center but refused to make the long airplane trip.

He was afraid of flying. ❂

NOTE: Like many circus histories, accounts of the origins of human cannonball acts are varied and subject to debate. The above article is based on several corroborating resources.

CHAPTER 20

STEAM CAR TECHNOLOGY: Built for the Fast Lane

If everything is coming
your way . . . you're in
the wrong lane.
– UNKNOWN

IT SEEMED LIKE A GOOD IDEA at the time: using steam to power early generations of the horseless carriage.

Yet in the 1930s, the steam car hissed quietly into the sunset of good intentions when the public's desire for convenience overcame its penchant for performance. Gone but not forgotten were thoughts of steam locomotion as a viable alternative to some of the troublesome energy and pollution issues long associated with internal combustion engines.

Although technology in the late 1800s focused heavily on moving the masses via locomotives and steamboats, steam cars opened the way for individual transport.

Compared to locomotives and steamboats of that era, steam automobiles experienced relatively few explosions. But occur they did, primarily during the experimental era of boiler design. The rarity of these explosions

The two-passenger steam car Locomobile, circa 1900, was equipped with a rubber bucket, side lamps, gong, cyclometer, a full set of tools, and a purchase price of $750.

was most assuredly the consequence of closer personal attention required by the driver to ensure his vehicle's continued operation. Indeed, had similar attention (i.e., proper maintenance and operation) been paid to other pressure-containing equipment, the frequency of pressure vessel accidents involving deaths and injuries might have been significantly mitigated.

Make no mistake: Steam car boilers generated massive amounts of power immediately available at acceleration. A two-cylinder double acting steam engine, like the Stanley, delivered as many power strokes as a V-8 internal combustion engine. Or consider a late model Dodge Viper: It is powered by a V-10 engine that produces 488 lbs./ft. of torque at 3,600 rpm. The simple two-cylinder, 20-hp Stanley can produce 593 lbs./ft. torque from rest.

Although never achieving a particularly endearing or enduring legacy, the steam car has not been without admirers. Comedian Jay Leno owns two he counts as among the favorites in his extensive car collection. Howard Hughes also had two, one of which—a Doble Model E—he took with him when he moved from Houston to California, leaving behind his Cadillac. The Stanley Steamer has been rated among the top 10 cars of the century by both *Road and Track* and *Car Collector* magazines. Many car buffs are surprised to learn the very first official presidential vehicle, used to chauffeur President William Howard Taft in 1909, was a steam car manufactured by White.

IN THE BEGINNING

Steam technology was around long before internal combustion, which first appeared about 1880.

Some believe the first steam car was developed in China by a Jesuit priest, Father Verbiest, in the mid-1600s. Sir Isaac Newton also tried his hand at a steam carriage in England around 1680. And the first automobile patent, granted to Oliver Evans in 1789, was for a steam car.

Early pioneers of steam technology during the eighteenth century included Frenchman Nicholas-Joseph Cugnot in the 1760s, and later Nathan Read of Massachusetts in the 1790s. And there were a number of pioneers during the nineteenth century, such as New

England's Sylvester Roper, who built steam cars and steam bikes from 1863 until his death in 1896 while riding his own steam bike at the Charles River Velodrome in Cambridge, Massachusetts.

Although Ransom E. Olds is best known for his gas-powered Oldsmobile, he built a steam car in 1887 and again in 1891 before switching to gas. In Boston at the turn of the nineteenth century there were George Eli Whitney and George Long working on steam car technology. And in the early years of the twentieth century, there were the White family, Abner Doble, and the Stanley twins.

ENTER THE STANLEYS

Francis Edgar (F.E.) and Freelan Oscar Stanley were doing quite well in their photography business, located in Kingfield, Maine, during the latter years of the nineteenth century. Credited with creating the airbrush and dry plate coating machine, they had a natural inclination for inventing things.

But the Stanleys' involvement with steam cars appears to have been born out of an incident involving F.E.'s wife. Having never learned to ride a bicycle, she crashed into a tree during her only effort. This prompted her husband to declare he would invent something they could both ride in.

Undoubtedly influenced by George Eli Whitney's 1896 steamer, F.E. decided to create his own motor carriage propelled by steam. He developed a number of prototypes before being invited to demonstrate his new invention as a novelty attraction at an 1898 Cambridge, Massachusetts, speed and hill-climbing demonstration.

There, his newly created carriage set an unofficial world record for the mile at a speed of slightly more than 27 mph. Curiously, none of the gas-powered cars at the event made it more than two-thirds the way up the 30-percent-grade hill. So it wasn't surprising no one took F.E. seriously when he attempted to round up a group of men to catch him and his vehicle when it was his turn to climb the hill ramp. Virtually shooting up to the top of the hill, F.E. was only able to stop by holding

Stanley Steamer 1910 Model 70.

open the throttle as spectators rushed to grab and subsequently contain the car.

Stanleys—even Whites early on—were virtually unbeatable during the twentieth century's first decade. This led to considerable consternation, frustration, even disgust, on the part of internal combustion purists. Whether it was a hill climb or a short-distance straight run, nothing outperformed steam cars, except perhaps a vehicle that benefited from considerable resources or extraordinary conditions. One such car was the special 60-hp, $18,000 (about $431,000 today) Mercedes imported for the Mt. Washington Climb to the Clouds race in 1904. It relegated the lowly 6.5-hp $650 (about $15,600 today) Stanley to second place in a field of more than 30 cars of which fewer than 20 actually finished.

When the Stanley Racer crashed in 1907, the Stanleys decided to quit competitive racing. They didn't want to endanger the lives of their employees with what they considered a frivolous activity. F.E. Stanley saw racing as having only marginal marketing benefit, and even less research and development value.

The "Standard Boiler" manufactured by the Steam Carriage Boiler Company was constructed of a seamless steel shell into which the crown sheet was riveted. It occupied the least amount of space and was easily fitted to any burner.

HOW DID STEAM CARS WORK?

Quite simply, a boiler was fired by pressurized pilot fuel igniting pressurized burner fuel to create steam released to a motor that in turn drove the rear axle.

Burners, boiler, and motor were connected through valves and mechanisms directly managed by the driver. With this form of external combustion, the driver was, in essence, controlling the engineering—that is, physically managing connection of the externally related components. (With internal combustion, engineering is actually built into the engine with the driver only having to turn the starter, steer the wheel, and pay attention to the rules of the road.)

Mechanically speaking, however, a steam car was comparatively elementary because it had few moving parts. The 20-hp Stanley engine, for example, had only 13 moving parts. To equal its power would require an eight-cylinder gasoline engine with 70 moving parts.

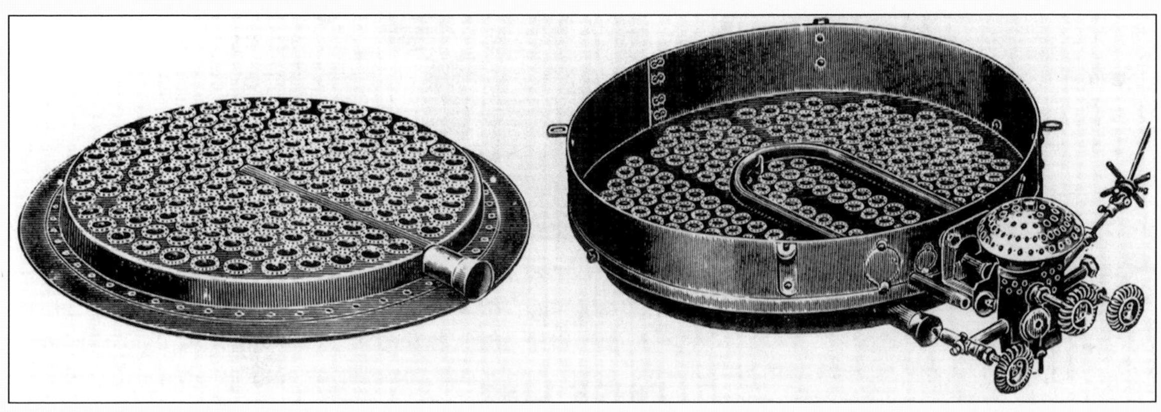

The "Reverse Burner Complete" featured an asbestos-lined casing, pilot light, generator, and regulator.

Mechanical simplicity notwithstanding, steam cars were complex to drive. Pressures, liquid levels, and lubrication were managed by the driver in the early model years of the Stanley. While the steam engine is easy to understand but difficult to operate, internal combustion is difficult to understand but easy to operate.

To convert thermal energy into mechanical motion, fire-tube boilers were used by the makers of Stanley, Locomobile, and Mobile. White and Doble makers employed water tubes or forced-circulation monotube steam generators.

Condensers were particularly important to steamers because they recovered expelled steam from the engine and converted it back to water. Simply explained: Steam is cycled through a radiator at the front of the car for cooling by the air flow. It is then returned to the water tank through a pipe, consequently extending a car's range of distance.

Mileage per gallon of water ranged from one to three miles. Under heavy use, the formula was reversed to gallons per mile, or gpm vs. mpg. Actual mileage varied by factors of engine age, weight, horsepower, and terrain. Generally, a non-condensing model would get one mpg of water. A condensing car would get approximately three.

At an average of one mpg of water, starting with a full boiler and a full water tank, water was replenished every 20 to 75 miles. Variables included condensing vs. non-condensing processes and size of the water storage tanks, which ranged from 20 gallons in the early horseless carriages to as high as 50 gallons in the largest 30-hp seven-passenger non-condensing touring cars.

And then there was the issue of maintenance. Yes, the steam engine was low-tech but maintenance was frequently necessary, messy, and labor intensive.

Citing the Stanley as an example: It required an estimated six to 12 hours of tinkering for every hour of driving. On one hand,

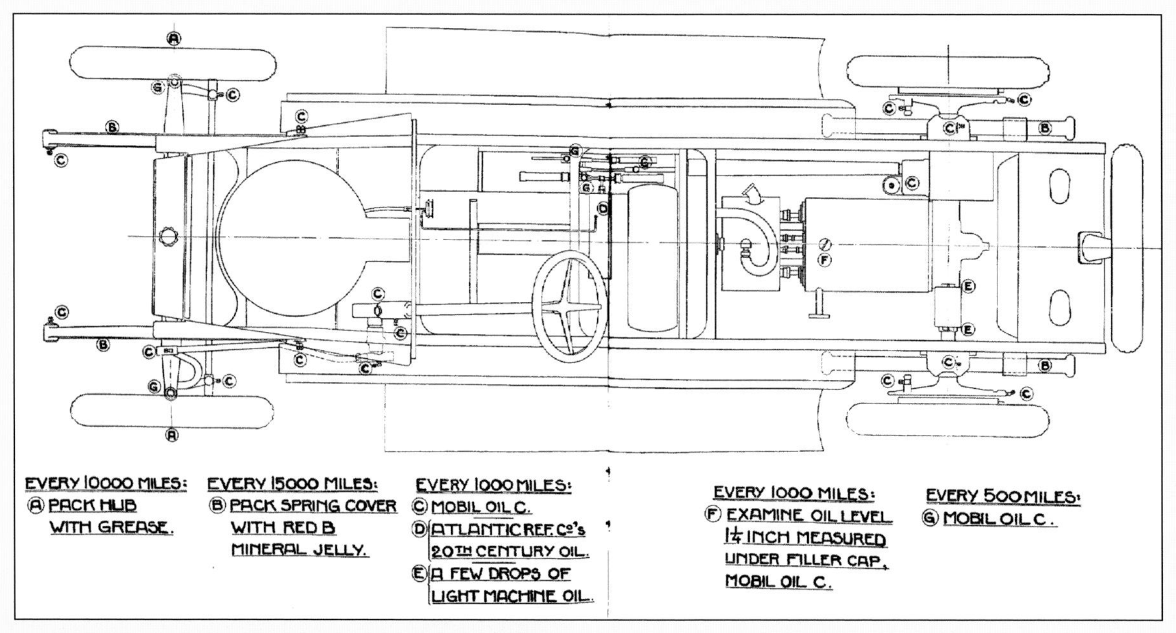

Stanley Steamer car maintenance schedule and guide.

many steam cars were so simple maintenance could be performed with a basic tool kit by anyone with minimal mechanical ability. But inattention to water or oil lubrication would put a steam car out of commission, just as running out of gas or poor spark plugs would disable a gas-powered vehicle.

Steam cars had a considerable amount of plumbing, including several sets of pumps for fuel and water supply. Then there was the boiler to generate steam and the burner to generate the heat to produce that steam.

All required continual maintenance.

TIGHT TORQUE

Torque is a measure of "twisting force." That is, the ability to twist an axle and turn a wheel and tire. The rate at which torque is applied is called "horsepower." Horsepower in a steam automobile is properly determined by the size of the boiler and then by the steam volume and pressure.

Naturally, steam car travel was about getting to where one wanted to go. Quickly. And, as mentioned, nothing was quicker than having maximum power at initial acceleration.

Top cruising speed for some 10 and 20 hp steam-driven models was 25 to 60 mph. For some 30 hp cars, that speed climbed as high as 100-plus mph. In 1906, the special Stanley "Rocket Racer" set the land speed record at 127.659 mph at Ormond Beach, Florida—on bicycle-tire technology. That record remained intact for four years until it was surpassed—by only four mph—by Barney Oldfield in the "Blitzen Benz."

Unlike internal combustion engines, which depend on fuel explosions, steam cars were virtually silent, with the exception of a low hiss of the burner and sound of rubber tires moving across pavement.

Of course, without a clutch and transmission, Stanleys and Dobles were somewhat simpler to handle than gas cars of the day. As for steering and maneuverability, all cars of that era were difficult to handle in proportion to their weight and size.

But steam cars didn't need to be cranked.

START YOUR ENGINES

It's interesting to note the Stanley was marketed as a perfect car for women since it could be "started with a match."

Because the boiler had to be fired and the steam generated, it typically took 20 minutes to start a fire-tube Stanley. (Flash generators in White and Doble steamers took as little as 30 seconds.)

Perhaps at the beginning of the day, 20 minutes might seem odious. But if the pilot burner had been left on overnight—a common practice among some Stanley owner-drivers—start-up time was nothing by comparison. And

Guttenberg, New Jersey, September 18, 1900: Locomobile cruises to win the five-mile race for steam carriages.

once fired and ready to go, a steamer had no equal: The steam engine developed maximum power from rest. With full power available at just a touch of the throttle, torque resulting from a typical automotive steam engine was massive and provided tremendous acceleration.

EVOLUTION OF STYLE

Early on, both internal and external combustion cars strove for light weight in order to improve performance. Once locomotion itself had been accomplished, however, the need for stability, safety, and comfort contributed to increased car weight.

From 1911 on, steam cars started getting heavier: front doors and aluminum skins over the wood bodies were added, and wooden frame members added significant heft. And then came condensers and steel frames to support the extra weight.

By the company's last year, 1924, Stanleys were "modern" looking, long-wheel based, comfortable—and veritable sloths. The 1910 Model 70 five-passenger touring car might have weighed 3,650 pounds loaded with passengers. The 1924 Model 750 five-passenger touring Stanley probably weighed closer to 5,000 pounds with passengers. By comparison, the legendary Doble Model E from the 1920s was powered by a massive engine, boiler, and burner that easily tipped the scale at 6,000 pounds.

Style notwithstanding, the first part of the car usually requiring attention, particularly in earlier models, was the wooden body. Rough riding over low-quality roadways of earlier eras could literally break a car to pieces.

Yet, the steam engine, when properly maintained, was virtually indestructible.

MARKETING THE STEAM CAR

There were more than 130 steam car manufacturers between 1896 and 1931 (generally considered the height of popularity for steam cars in North America). Both White and Stanley cars were commercially successful in the first decade of the century—at least they made enough money to justify manufacture.

All steam vehicles were hand-assembled. The Whites manufactured nearly 11,000 vehicles between 1900 and 1910. Stanley manufactured the same quantity between 1900 and 1924. Locomobile manufactured more than 5,000 cars from 1899 to 1904. The Doble line generated more than 5,000 orders following a New York car show in 1917 at an announced price of $2,500 (about $42,000 today) per unit. But the company went out of business shortly thereafter. Doble blamed its problems on being unable to obtain steel following the United States' entry into World War I. Others suggested difficulties within the Doble family and that the company may have had severe technical problems with the car itself.

The years 1896 to 1906 saw the most interest in steam cars. These vehicles still enjoyed a following into the 1910s even though they were more despised than popular, especially by internal combustion purists.

The advantage of internal combustion operation was enhanced by the electric starter in 1912.

As for a steam car's appeal to the general public: Manufacturers made a strategic marketing error in trying to engage the rare driver who had a genuine appreciation for superior performance.

The general public just wanted to get in a car and go. ❂

THE SHORT AND UNEVENTFUL HISTORY OF STEAM CAR BOILER EXPLOSIONS

Dave Nergaard has been collecting data on automobile steam technology part-time for more than 40 years.

His research as vice president of the Steam Automobile Club of America reveals there were more than 55,000 steam cars manufactured and sold in the United States (most before 1905). Of all the statistics Nergaard has cobbled together, one confounds him:

"I have . . . found remarkably few references to automotive boiler failures, and only 10 mentions of explosions with any detail. It is surprising that there were so few accidents!"

ITEM: *"In 1834, a Scott Russell coach broke a wheel trying to negotiate a pile of rocks deliberately placed to block the road. The wheel failure caused the boiler to assume loads previously carried by the coach chassis and axles, which it was unable to do. The resulting explosion killed several people and ended the commercial viability of those coaches. However, the out-of-court settlement of the suit brought by Scott Russell against the turnpike company essentially absolved that boiler from fault."*

ITEM: *"[A] Walter Hancock coach exploded when the engine man tied down the safety valve and ran the declutched engine so the fan would force the fire! I don't think I have to say much about this failure, and neither did the coroner's court . . ."*

ITEM: *"A blacksmith obtained a boiler rejected and scrapped by the Stanley Company and attempted to use it without the usual wire wrapping used to strengthen Stanley boilers. According to the story, he destroyed the boiler, his shop, and himself. Thereafter, the Stanley factory drove a spike through the side of any rejected boiler shell."*

ITEM: *"August 10, 1908, near Painesville, Ohio, two people killed. The boiler . . . was not the boiler normally fitted to the car. . . . The accident happened near the bottom of a long, downhill grade, and it is assumed that the fire had not shut off when it should have and the safety valve had stuck."*

Nergaard reports no explosions have occurred since 1910.

"There have been steam car explosions in recent years," he concludes, "but all involved fuel tanks, not boilers. There have been boiler failures also, but these have not been explosions."

The steam car of today holds the wheel driven land speed record of 139.843 mph. Developed by the British Steam Car Challenger Team, the Inspiration *eclipsed the 103-year land speed record of 127.659 mph established by the Stanley Steamer in 1906.* Inspiration *was nicknamed the "Fastest Kettle in the World" because the burners can produce 23 cups of tea per second. Containing over two miles of tubing, it is powered by 12 stainless steel suitcase-sized boilers fueled by liquid petroleum gas. Pressurized at 40 bar, water is pumped into the boilers at about 11 gallons per minute. Superheated steam—at twice the speed of sound—enters a two-stage 13,000 rpm turbine that drive the rear wheels.*

CHAPTER 21

STEAM: Whence a San Francisco Tradition

It isn't the mountain ahead
that wears you out; it's the
grain of sand in your shoe.
– *Robert W. Service, English-born poet, 1874–1958*

San Franciscans may not know it, but they owe a great deal of gratitude to steam power. And they need only look as far as the street for the reason why.

The city's cable car system is as unique as it is functional. Over a period of nearly 140 years, it has become a charming, readily identifiable symbol of the City by the Bay.

San Francisco's hilly terrain has always been somewhat tricky to navigate, even given today's modern applied science of transportation. But the practicality of cable car technology is deeply rooted in the world's first cable car line: the Clay Street Hill Railroad, established August 2, 1873.

Clay Street was a notoriously precipitous avenue even by San Francisco standards. It has been posited the cable system was born out of the city's concern for horses pulling heavy streetcars up its steep streets.

The first cable car on Clay Street at Kearny photographed in 1873. Bettmann/Corbis/AP Images.

The Clay Street Hill Railroad was the first to employ a grip process which enabled the car to "grip" a cable traveling a preset route. The source powering the cable: a stationary steam engine.

The railroad was absorbed by the Ferries & Cliff House Railway Company in 1888 as the now famous Powell Street route began service. The prevailing technology of the industrial era, stationary steam engines were instrumental in the evolutionary success of the cable car phenomenon. Each engine was powered by anthracite coal, the smoke from which spewed from a 185-foot newly constructed smokestack.

Van Ness Ave., California
& Market Streets

SAN FRANCISCO CABLE CAR FAST FACTS

- Each cable car grip is comparable to a 260-pound pair of pliers.
- Cable speed is a constant 9.5 mph.
- Pressure on the cable is as much as 30,000 pounds per square inch.
- A cable generally stretches several feet during its period of service.
- While the grip provides locomotion, it is relaxed when the cable car approaches a designated stop, at which time the conductor applies the vehicle's brake system.

The first cables needed to move two route lines of cars were driven by two 450-horsepower, high-pressure horizontal slide valve, Corliss-type Thompson non-condensing, stationary steam engines. Popularity of the cable cars in 1890 prompted the railway company to construct two additional Corliss-type stationary engines each rated 500-horsepower. Cost of each unit: between $18,000 and $20,000 (about $431,000 and $479,000 today).

In 1894, the original boilers were replaced with four batteries, each consisting of two elephant boilers (box-shaped water tube units as opposed to cylindrical). With an individual horsepower of 920, the boilers were fed coal delivered by horse-drawn drayage.

Coal consumed daily by the railroad company was approximately ten tons. In 1893, it cost $1.50 (about $36 today) per car to operate an average fifty-one-car fleet (price of coal: $8 per ton or about $200 today). The thick black smoke and soot billowing from the smokestack—as well as the increasing cost of coal—prompted in 1901 a change to oil as the cable system's fuel source.

In 1912, for purposes of economy and reliability, electric motors replaced most of the steam engines. However, steam was used intermittently as backup power until 1926. Today, four cables are each driven by 510-horsepower DC electric motors.

With all evidence of cable cars' flirtation with steam long gone, there remains but one connection to an era of inspiration and incubation.

The 185-foot smokestack, now reduced to 60 feet, remains proudly erect at Washington and Mason streets as the last remnant of the 1880s and the first steam-powered cable cars. ☼

CHAPTER 22

MOLASSES MADNESS: The Purity Distilling Company Explosion of 1919

Reality is merely an illusion...
albeit a very persistent one.
– Albert Einstein, German-born American physicist, 1879–1955

It was perhaps the most bizarre pressure equipment accident in North American history.

The place: Boston's Commercial Street Wharf. *The date:* January 15, 1919. *The location:* Purity Distilling Company.

January 15 was unlike other January days in Bean Town. Just three days before, temperatures dropped to two degrees F. But on this mid-winter day, temperatures climbed into the 40s.

As always, the Commercial Street Wharf area was alive with commerce. Densely peopled with poor Irish and Italian laborers, streets of the North End accommodated endlessly chugging motor trucks en route to area businesses and docked ships. Horse-drawn wagons hobbled the pavement making scheduled deliveries. Trolleys filled with passengers rolled to and fro on elevated tracks. And directly across from the wharf, 65-year-old Bridget Clougherty carefully suspended her laundry across a porch clothesline.

During the early 1900s, molasses was a particularly desirable commodity. Used in the production of rum and food, it was also employed to make industrial alcohol for the production of ammunition.

And so it was the Purity Distilling Company sought to maximize output of the gooey confection.

A cast iron tank constructed in 1915, measuring 50 feet high and 90 feet in diameter, had a capacity to hold over 2.5 million gallons of crude molasses.

While production workers at the distillery basked in spring-like weather during their lunch period, trouble was brewing—quite literally—nearby.

It was approximately 12:40 P.M. when the laborers sensed movement below their feet accompanied by a low, reverberating rumble. Almost instantly, it was followed by the pressure wave of an explosion.

As reported by the Boston *Evening Globe*:

> Fragments of the great tank were thrown into the air, buildings in the neighborhood began to crumble up as though the underpinnings had been pulled away from under them, and scores of people in the various buildings were buried in the ruins, some dead and others badly injured.
>
> The explosion came without the slightest warning. The workmen were at their noontime meal, some eating in the building or just outside, and many of the men in the Department of Public Works buildings and stables, which are close by, and where many were injured badly, were away at lunch.

While the explosion was deadly beyond a doubt, it set in motion a series of events that can scarcely be imagined. According to the *New York Times*:

> A dull, muffled roar gave but an instant's warning before the top of the tank was blown into the air. The circular wall broke into two great segments of sheet iron which were pulled in different directions. Two million gallons of molasses rushed over the streets and converted into a sticky mass the wreckage of several small buildings which had been smashed by the force of the explosion.

Reports from that day indicated the molasses created waves as high as 15 feet. As the thick, sweet goo cascaded onto the street, it took with it horses, buildings, cars, wagons, and people. A firehouse close by was lifted from its foundation. For a two-block area, those attempting to flee the syrupy mess were quickly overcome by a torrent traveling an estimated 35 miles per hour. Some were tossed into stationary objects. Many simply drowned.

The next day, the Boston Post wrote:

> The sight that greeted the first of the rescuers on the scene is almost indescribable in words. Molasses, waist deep, covered the street and swirled and bubbled about the wreckage. Here and there struggled a form—whether it was animal or human being was impossible to tell. Only an upheaval, a thrashing about in the sticky mess, showed where any life was . . . Horses died like so many flies on sticky fly paper. The more they struggled, the deeper in the mess they were ensnared. Human beings—men and women—suffered likewise.

Rescuers found the task of retrieving bodies particularly difficult. Moving around in the dirty glop often resulted in boots being swallowed directly from the feet of the volunteers. Stuck horses had to be shot. It took months for cellars around the explosion site to be siphoned free of the mucky substance.

Streets and homes were later cleaned with salt water. Although months passed, those walking about the Commercial Street area often felt their shoes sticking to pavement. (To this day, some North End residents swear the sweet aroma of molasses remains, particularly on hot afternoons.)

Human deaths numbered 21, including Miss Clougherty. A total of 150 were injured. Additionally, a significant number of horses were killed. Property damage amounted to almost $100 million in today's dollars.

The days following this horrendous tragedy saw a number of theories as to why the tank exploded. Among the most curious was a charge by the company owner (U.S. Industrial Alcohol) that anarchists sabotaged the tank with dynamite. (Boston was a hotbed for anarchists at the time.)

Unlike the company, families of the poor laborers had few resources. However, they were able to prove the tank was of shoddy construction, citing numerous leaks that had been caulked or patched. Since tank erection four years earlier, leakage was so frequent the company attempted to conceal the seepage by painting the huge structure brown.

A court-appointed military officer found in the families' favor, citing no credible evidence of sabotage. In 1925, the company was ordered to pay $1 million ($13.1 million today) to settle 119 separate suits.

Although this tragic explosion involved substandard construction four years earlier, what actually shattered the tank was probably a circumstance few would have foreseen.

It is believed the structure was filled almost to its 2.5 million gallon capacity. As the warm molasses poured into the five-story tank, it mixed with cool atmospheric temperatures. The result activated a fermentation process, which in turn generated gas that created pressure leading to the tank's destruction.

What puzzled investigators assigned to the explosion was a lack of information on the massive container. Sourcing every state and local agency involved with construction, it was learned confusion of the terms *industrial device* and *structure* prompted two responsible agencies each to assume the other had taken the appropriate measures to certify construction.

BOTTOM LINE: There were no building plans. Consequently, no building plans had been filed or approved. Nor had the tank ever been visited by government inspectors.

SCENES FROM THE PURITY DISTILLING COMPANY EXPLOSION OF 1919

All photographs by Leslie Jones, January 15, 1919. Images provided courtesy of the Boston Public Library, Print Department.

Damage from the explosion was estimated to be $100,000,000 (2008 dollars). Reports indicated the syrupy goo moved at 35 mph (56 km/h), and exerted a pressure of 2 tons/ft^2 (200 kPa).

A large portion of the tank remained intact after the explosion. Witnesses would later report that as the tank collapsed, a machine-gun-like sound could be heard as rivets shot out of the metal-containing walls. Rumors the company was overstocking molasses for conversion to rum in anticipation of Prohibition were later discounted.

View of Engine 31 Fireboat house, ripped from its foundation following the explosion. When the second floor collapsed onto the first, it trapped several firefighters, one of whom was George Layhe. Face-up with legs crushed under the weight of a piano and a pool table, Layhe managed to keep his head above the rising molasses for several hours. Finally exhausted, he relaxed his head backward and drowned.

Firemen attempted to rescue their fellow firefighting brothers at Engine 31 Fireboat house. Efforts were halted after four days, with many of the victims covered in molasses and difficult to recognize. It was estimated cleanup efforts consumed over 87,000 man-hours.

Physically devastating as the aftermath appeared, perhaps nothing was as shocking as observing twisted steel girders supporting the overhead train system. Those close to where the tank once stood reported a sudden vacuum creating a reverse shockwave that consumed the air, while at the same time knocking down every person and object in the immediate area.

CHAPTER 23

TAPPED OUT: The London Beer Flood of 1814

Every path hath a puddle.
– *George Herbert, Welsh poet, orator, priest, educator, 1593–1633*
Jacula Prudentum *(1640)*

It was hard to miss at the junction of London's Tottenham Court Road and Oxford Street.

The Horseshoe Brewery was known for the huge vats of beer standing tall atop its 25-foot high building. But it was an intimidating 22-foot-tall cask containing the company's porter beer that really commanded one's eye.

On April 1, 1785, *The Times* described the mammoth vat thusly: "The size … exceeds all credibility, being designed to hold 20,000 barrels of porter; the whole expense attending the same will be upwards of £10,000 (approximately $9 million today)."

Constrained by 29 iron hoops, the wooden vat for years admirably served in holding 135,000 imperial gallons of brew. That is, until October 17, 1814.

The St. Giles Rookery in London.

Brewery workers may or may not have noticed the vat showing stress consistent with nearly 30 years of containing a massive quantity of porter.

About six P.M., one of the hoops blew apart with devastating concussion as the vat violently exploded, spilling beer, metal and pieces of wood across the roof. Impact was immediate. A chain reaction toppled the other vats, setting in motion the release of even more of the brewery's intoxicating product. The ear-shattering blast could be heard as far as five miles away.

A total of 323,000 imperial gallons of beer poured into the brewery building and the surrounding slums of St.Giles. While some of the residents might momentarily have perceived their prayers granted by a higher authority, they soon realized the waves of beer flowing down the streets would take a devastating toll on nearby tenements and overcrowded basements of the slum dwellings.

Within seconds, the cascade of brew demolished two houses before unleashing its devastation upon the Tavistock Arms pub on Great Russell Street. Here, 14-year-old barmaid Eleanor Cooper became trapped while watching in horror as the newly fermented beverage reduced a pub wall to rubble.

In an attempt to take advantage of their good fortune, hundreds of residents attempted to scoop up the beer with kettles and pots. Others simply lowered themselves to the street and opened their mouths. While perhaps a good idea in theory, the latter plan was quickly abandoned by many when the surge proved to be more than they could consume.

As the injured arrived at Middlesex Hospital, a small but lively skirmish ensued when patients already in the hospital detected a strong odor of beer wafting about the flood survivors. Quickly discerning flood victims had participated in a beer party to which they were not invited, the patients demanded their fair share of the foamy refreshment.

As the wave of beer began to recede from St. Giles' streets, residents came to understand the reality: their homes, property, and lives in despair.

Mary Mulvey and her three-year-old son Thomas were among those who drowned. Hannah Banfield and Sara Bates, four and three years old respectively, were swept away by the tide and later died of injuries they sustained.

Family members of the deceased, desperate in their poverty, arranged for loved ones to be laid out at home where they charged mourners a fee to view the corpses. Visitation at one victim's house was so heavy that the floor collapsed, dropping curious onlookers into a basement still half-filled with heady brew. After police shut down the ghoulish entrepreneurs, St. Giles residents contributed to funeral costs by placing coins on the victims' caskets.

As expected, the stench of stale beer endured for months despite cleanup efforts. Also damaged beyond repair were many of the modest houses and livelihoods literally washed away by the unusual tidal wave.

Those adversely affected by the flood took Meux, owner of the Horseshoe Brewery, to court. However, a judge ruled the calamity an "act of God," meaning no person or entity was responsible.

Site of the Horseshoe Brewery was demolished in 1922 and is today partially occupied by the Dominion Theatre.

Nine deaths were attributed to drowning and injuries sustained from the flood.

One person died of alcohol poisoning. ❂

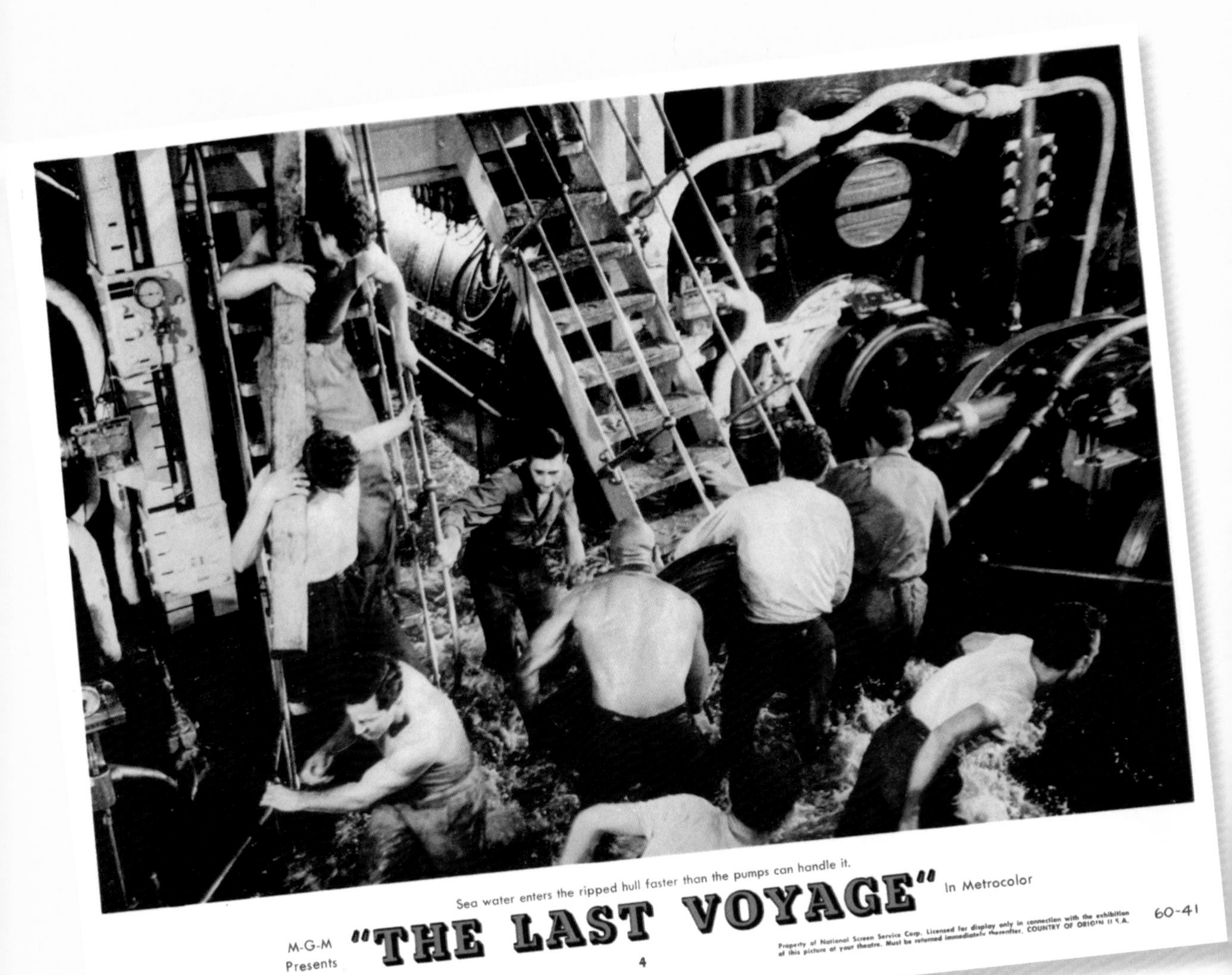
Sea water enters the ripped hull faster than the pumps can handle it.
M-G-M Presents
"THE LAST VOYAGE"
In Metrocolor
60-41
4
Property of National Screen Service Corp. Licensed for display only in connection with the exhibition of this picture at your theatre. Must be returned immediately thereafter. COUNTRY OF ORIGIN U.S.A.
Copyright © 1960 Loew's Incorporated

CHAPTER 24

NOTHING SAYS LOVE LIKE A BOILER EXPLOSION

Smooth seas do not
make skillful sailors.
– *African Proverb*

Although it may seem a somewhat disparate association, there exists a movie best described as a boiler explosion love story.

And here's what's surprising: As cinema goes, it makes for a pretty good adventure.

The film begins with a cruise ship captain at his table entertaining passengers during lunch. Without as much as establishing the movie's premise, he is handed a note that simply reveals: "Fire in the engine room."

And so commences a Technicolor thriller dealing with a catastrophic boiler accident aboard an aging luxury ocean liner.

Released in 1960, *Last Voyage* tells the story of a narcissistic ship's captain and his failed efforts to keep afloat the fictional SS *Claridon* following an explosion that rips a vertical elevator-sized hole through the vessel's mid-section. But the main plot focuses on Cliff and Laurie Henderson (Robert Stack and Dorothy Malone), a married couple on a cruise en route to Japan with their precocious daughter Jill (Tammy Marihugh).

As the story unfolds in the mid-Pacific, the engine room fire causes boiler controls and safety valves to fuse shut, thus causing dangerous internal pressures.

What makes this movie different from boiler room sequences in other productions is the number of scenes devoted to preventing an explosion and consequently the ship from sinking.

Suffice it to say the boiler does explode, and Laurie Henderson becomes trapped in her stateroom by a twisted piece of iron bulkhead that precludes her exit along with other passengers. The last 30 minutes of the picture focus on the efforts of her husband and the liner's engineering crew to free the woman before water fills her cabin and eventually the ship.

The boiler room scenes are exceptionally well-done and illustrate vividly what can happen in an emergency situation. Anxiety is compounded after the explosion when a boiler tender (stoker) is sent to find the ship's chief engineer, who attempted to dislodge a stuck safety valve.

According to the returning tender, "There isn't much of the chief to scrape up. . . ."

By comparison with other disaster films, this production is stark, with an ending as abrupt as its beginning. The only background music occurs during the introduction and closing credits. In what could be considered a first for films of that genre, the events aboard ship appear to take place within a real-time sequence of 91 minutes.

Another creative cinematic rarity is reflected in the use of a real cruise ship to film many of the onboard and exterior sequences, including the actual sinking scenes at the film's conclusion. For a production completed in 1959, the movie's special effects are well-accomplished even when compared to the technical wizardry and electronic standards of today. *Last Voyage*'s artistry earned it a nomination for a Best Visual Effects Academy Award.

On the subject of Academy Awards, *Last Voyage* boasted three Academy Award-winning actors (Malone; George Sanders, who played the captain; and Edmond O'Brien, who portrayed the second engineer). Male lead Robert Stack was an Academy Award nominee.

While *Last Voyage* may not have been the first disaster film, it was among a group of period movies that spawned—some say laid the foundation for—disaster extravaganzas of the future. It certainly was the first and only such production with a plot line focused on a boiler explosion.

Among the disaster epics to follow would be *Towering Inferno*, *The Poseidon Adventure*, *Airport,* and *Titanic,* to name several.

Over 50 years old, *Last Voyage* has held up surprisingly well and can still generate viewer angst. It can occasionally be seen on classic movie TV channels. ❂

SCENES FROM *THE LAST VOYAGE*

Most of interior and exterior scenes were shot in 1959 aboard the SS Ile de France, a retired cruise liner scheduled to be scrapped. It was reported all boiler scenes were actually filmed in the liner's engine room.

As with all movies, creative license was taken in a number of instances. Example: the midsection hole extending from the boiler room to the top of the ship most likely would have been caused by a piece of heavy equipment, presumably a boiler. None was visually displayed or referenced in the film.

The Last Voyage *was based upon a real incident resulting from the* Andrea Doria *tragedy off the coast of Cape Cod in July 1956. Martha Peterson was trapped by wreckage in her cabin and eventually drowned hours before the* Andrea Doria *sank. Ironically, the* Ile de France *was one of several boats that came to the* Andrea Doria*'s aid.*

"We're only sinking; let's not alarm the passengers by telling them."
– Captain Robert Adams (played by George Sanders)

EPILOGUE

I could have saved a life that day,
But I chose to look the other way.
It wasn't that I didn't care;
I had the time, and I was there.

But I didn't want to seem a fool,
Or argue over a safety rule.
I knew he'd done the job before;
If I spoke up he might get sore.

The chances didn't seem that bad;
I'd done the same, he knew I had.
So I shook my head and walked by;
He knew the risks as well as I.

He took the chance, I closed an eye;
And with that act, I let him die.
I could have saved a life that day,
But I chose to look the other way.

Now every time I see his wife,
I know I should have saved his life.
That guilt is something I must bear;
But isn't' something you need share.

If you see a risk that others take
That puts their health or life at stake,
The question asked, or thing you say;
Could help them live another day.

If you see a risk and walk away,
Then hope you never have to say,
"I could have saved a life that day,
But I chose to look the other way."

– DON MERRELL

ABOUT THE AUTHOR

As Director of Public Affairs for The National Board of Boiler and Pressure Vessel Inspectors, Paul Brennan Jr. is responsible for all communications, marketing, and governmental affairs functions.

Graduated with B.A.s in Journalism and English from Kent State University, Mr. Brennan began his career in northeast Ohio as a broadcast journalist. As such, he reported numerous news stories for major national broadcast networks and wire services.

The award-winning author began writing professionally at the age of 19. He has penned a number of articles and lectured on the subjects of marketing and governmental affairs at universities and before international professional groups and associations.

During a career spanning 45 years, the Pittsburgh native has been honored with both the Gold Quill and the Silver Anvil. He is a former professional advisor to journalism students at The Ohio State University.

THE NATIONAL BOARD OF BOILER AND PRESSURE VESSEL INSPECTORS

SINCE 1919, the not-for-profit National Board has been composed of chief boiler and pressure vessel inspectors representing states, cities, and provinces enforcing pressure equipment laws and regulations. Created to prevent death, injury and destruction, these laws and regulations represent the collective input of National Board members.

For the general public, the importance of thoroughly trained and specially commissioned inspectors is of critical significance: Every person in the civilized world comes within close proximity of pressure equipment several times each day.

To advance the vital cause of safety, the National Board provides professional services of noteworthy importance to the pressure equipment industry:

National Board Training

Every year the National Board hosts hundreds of boiler and pressure equipment professionals from around the world at its training facilities, located on an attractive 16-acre wooded campus in Columbus, Ohio. National Board training reflects its credibility and reputation as a third party evolving from the regulation and enforcement side of the boiler and pressure vessel industry.

National Board Registration

Registering a pressure-retaining item with the National Board requires certain uniform quality standards be achieved certifying the manufacturing, testing, and inspection process. This certification acknowledges to owners, users, and public safety jurisdictional authorities that registered items have been inspected by National Board-commissioned inspectors and built to required standards.

Registration is required by most U.S. jurisdictions for installation of pressure equipment.

Since the process began in 1921, over 45 million pressure items have been registered with the National Board.

National Board Pressure Relief Department and Testing Laboratory

Each year, representatives from around the world travel to the National Board Testing Laboratory north of Columbus, Ohio. The purpose: to accurately measure the performance of their companies' pressure relieving devices.

Tested products undergo independent certification of function and capacity.

Capacity certification signifies equipment designs have been thoroughly reviewed. Additionally, it indicates the quality system has been audited and the equipment meets internationally recognized standards for preventing potential overpressure conditions in boilers and pressure vessels. Testing is also performed to evaluate a company's ability to properly repair pressure relief valves.

The National Board lab supports industry research and development by testing new designs, serving as a comparative standard for other laboratories, validating new concepts, and—upon jurisdiction request—assisting in boiler and pressure vessel incident investigations.

National Board Accreditation

The National Board administers three accreditation programs for organizations performing repairs and alterations. Accreditation involves a thorough evaluation of the organization's quality system manual, including a demonstration of its ability to implement the system. Authorized repair organizations are issued symbol stamps for application to equipment nameplates signifying the integrity of work performed.

National Board Inspection Code

As flagship publication of the National Board, the *National Board Inspection Code* (NBIC) is a consensus document created by an evolving committee of pressure equipment professionals.

Distributed biennially, the NBIC provides rules, information, and guidance to manufacturers, jurisdictions, inspectors, repair organizations, owner-users, installers, and contractors, as well as other individuals and organizations performing or involved in post-construction activities. THE OBJECTIVE: to provide uniform administration of rules pertaining to pressure equipment items.

The NBIC was first published in 1945 and is today the *only* standard recognized worldwide for inservice repair and alteration of boilers and pressure vessels. Approved as an American National Standard (ANSI) in August 1987, the *National Board Inspection Code* has been adopted by a number of states and jurisdictions, as well as federal regulatory agencies, including the U.S. Department of Transportation.

National Board BULLETIN

National Board's technical journal is distributed worldwide three times annually.

In addition to articles of interest to the pressure equipment industry, the *BULLETIN* provides helpful tips on equipment inspection, repairs and alterations; industry case histories; and a comprehensive listing of jurisdiction law and regulation amendments. Readers also find technical perspective from National Board staff and guest columnists, a complete listing of offerings from the training department, and the latest violations tracking data.

National Board Website

The National Board Website at *www.nationalboard.org* is the pressure equipment industry's premier information resource. Regularly updated twice monthly with topical pressure equipment news, it serves as a virtual library for those seeking accurate, comprehensive insight on a wide variety of important industry issues. Most information is free of charge.

Pressure equipment professionals from over 160 countries visiting the National Board site can look up laws and regulations for every North American jurisdiction, comment on proposed changes to the *National Board Inspection Code*, take an online training class, search a comprehensive up-to-date listing of manufacturers and repair organizations, and instantly search and read entire copies of past *BULLETIN* issues.

National Board General Meeting

Each spring, the General Meeting is conducted in conjunction with the American Society of Mechanical Engineers to address important issues relative to the safe installation, operation, maintenance, construction, repair, and inspection of boilers and pressure vessels.

Attendees include boiler and pressure vessel inspectors, mechanical engineers, engineering consultants, equipment manufacturers, representatives of repair organizations, operators, owners and users of boilers and pressure vessels, labor officials, welding professionals, insurance industry representatives, and government safety personnel.

General Session presentations cover a wide range of pressure equipment topics such as safe operation, maintenance and repair, safety valves—as well as other unit components—testing codes and standards, risks and reliability, and training.

National Board Scholarship Program

The National Board annually offers $6,000 scholarships to two outstanding college students meeting eligibility standards.

Scholarships are available to the children, step-children, grandchildren, or great-grandchildren of past or present National Board Commissioned Inspectors (living or deceased). These are also available to children of past or present National Board employees (living or deceased). ❁

SELECTED BIBLIOGRAPHY

THE FOLLOWING IS A LIST of selected works used by the author in the creation of ***BLOWBACK****. It is representative of the numerous wide-ranging references employed in development of the author's thoughts and ideas. Although by no means a complete record of resources consulted, this information is provided as a courtesy to those readers seeking a more thorough understanding of safety and the potential dangers associated with pressure-retaining equipment. Listed Web addresses were accessible at time of publication.*

"1974: Human Cannonball Misses Target," BBC News On This Day, August 25. *BBC News Online*, http://news.bbc.co.uk/onthisday/hi/dates/stories/august/25/newsid_2535000/2535601.stm (accessed Sept. 25, 2008).

"2 Sentenced for Party Blasting." *The Ironwood* (Mich.) *Daily Globe*, July 3, 1981: 13.

"Abraham Lincoln, Inventor." *Popular Mechanics*, March 1924: 360–363, Modern Mechanix blog, http://blog.modernmechanix.com/abraham-lincoln-inventor/ (accessed Oct. 14, 2009).

"Act: Thrill." *The Circus in America*, www.circusinamerica.org/public/acts/public_show/154, (accessed Sept. 22, 2008).

Adams, Cecil. "Can High-Pressure Steam Cut a Body in Half?" *The Straight Dope*, Aug. 4, 2006. http://www.straightdope.com/columns/read/2665/can-high-pressure-steam-cut-a-body-in-half (accessed Jan. 29, 2007).

Aldrich, Mark. *Death Rode the Rails: American Railroad Accidents and Safety, 1828–1965*. Johns Hopkins University Press, 2006.

American Society of Heating, Refrigerating and Air-Conditioning Engineers, Inc. *ASHRAE Handbook – Fundamentals, Inch-Pound Edition*. Atlanta: ASHRAE, 1997.

"Ask Dr. Baden: Does a Burn Caused by Steam Differ Physically from One Caused by a Flame?" *HBO.com*, 2008.

"Automatic Food Cooker Runs by Exhaust Heat of Car." *Modern Mechanix Hobbies and Inventions*, June 1930. via *Modern Mechanix* blog, April 18, 2008, http://blog.modern mechanix.com/2008/04/18/automatic-food-cooker-runs-by-exhaust-heat-of-car/ (accessed June 14, 2011).

Axtman, William. "A Boiler: The Explosive Potential of a Bomb." *National Board BULLETIN*, Winter 1996: 31–32.

"Beer Keg Explosion Kills Man." Associated Press. Frederick (Md.) *Post*. Oct. 31, 1988: A-3.

"Beer Keg Explosion Kills Man." Associated Press. *WCBS NewsRadio 880* (New York), Oct. 23, 2006.

"Beer Keg Explosion Kills Man." United Press International. *Tyrone* (PA) *Daily Herald*. Aug. 4, 1981: 8.

"Beer Keg in Bonfire Explodes." Associated Press. *Indiana* (Penn.) *Gazette*, May 1, 1995: 4.

"Beer Keg 'Takes off like a Rocket,' Killing Student." United Press International and Associated Press. *Pacific Stars and Stripes*. Aug. 6, 1981: 5.

Bellamy, John Stark II. "Streets of Hell," Chap. 1 in *Cleveland's Greatest Disasters!* Cleveland, Ohio: Gray & Co., 2009.

Bengston, Harlan. "The Invention of the Steam Engine," *Bright Hub Engineering*, July 26, 2010, www.brighthub.com/engineering/mechanical/articles/71480.aspx (accessed June 14, 2011).

Berry, Rev. Chester D. *Loss of the Sultana and Reminiscences of Survivors*. Lansing, Mich.: Darius D. Thorp, 1892.

"Big Rig Myths." *Mythbusters* TV science series, Episode 80, Discovery Channel, June 6, 2007. via *Mythbusters Results*, www.mythbustersresults.com/episode80 (accessed Apr. 9, 2010).

"Blast Wrecks House and Injures Suspected Thieves." *Birmingham* (UK) *Post*, Oct. 24, 2006, http://www.birminghampost.net/news/west-midlands-news/2006/10/24/blast-wrecks-house-and-injures-suspected-thieves-65233-17981971/ (accessed Feb. 13, 2012).

"Born to be Wild: Sylvester Roper, Inventor of the First Motorcycle." *National Board BULLETIN*, Fall 2007: 18–25.

"Boy Killed by Anal Penetrating Chair." *Anorak*, Feb. 20, 2009, http://www.anorak.co.uk/202541/strange-but-true/boy-killed-by-anal-penetrating-chair.html/ (accessed Sept. 13, 2011).

Brennan, Paul. "Pressure Equipment Accidents: Communicating the Human Toll." *National Board BULLETIN*, Summer 2008: 6–7.

____. When Is A Boiler Explosion Not A Boiler Explosion?" *National Board BULLETIN*, Summer 2007: 3–5.

Buckman, David Lear. "Disasters of River Travel," in *Old Steamboat Days on the Hudson River*, Grafton: 1907.

Buduson, Sarah. "Recent Propane Explosions Cause Concern." *newsnet5.com* (WEWS ABC-TV, Cleveland), Mar. 2, 2011, www.newsnet5.com/dpp/news/local_news/investigations/recent-propane-explosions-cause-concern (accessed Mar. 3, 2011).

Buer, Mabel C. *Health, Wealth and Population in the Early Days of the Industrial Revolution.* London: George Routledge & Sons, 1926.

Cannell, John Clucas (J.C.) *The Secrets of Houdini*. London: Courier Dover Publications, 1973.

Celizic, Mike. "Man Who Cut Off His Arm Was 'Convinced I'd Die.'" NBC *Today Show*, June 22, 2010, http://today.msnbc.msn.com/id/37832454/ns/today-today_people#slice-2 (accessed Oct. 15, 2010).

Cengel, Yunus A. and Robert H. Turner. *Fundamentals of Thermal-Fluid Sciences*. McGraw-Hill Higher Education, 2001: 74–75.

"Clay Street Hill Railroad — 1873." *San Francisco Cable Car Museum*, www.cablecarmuseum.org/co-clay-st-hill.html (accessed Nov. 17, 2008).

Cleveland, Cutler J. "Papin, Denis." *Encyclopedia of Earth*, Aug. 18, 2006, www.eoearth.org/article/Papin,_Denis (accessed Aug. 20, 2012).

"The Cleveland East Ohio Gas Explosion of 1944." *All Things Cleveland Ohio*, Dec. 5, 2008, http://allthingsclevelandohio.blogspot.com/2008/12/cleveland-east-ohio-gas-explosion-of.html (accessed Aug. 20, 2012).

Cole, Jean. "Man Seriously Injured When Moonshine Still Explodes." *The News-Courier* (Athens, Ala.), Sept. 30, 2008, www.enewscourier.com/x1037419052/Man-seriously-injured-when-still-explodes (accessed Sept. 30, 2009).

"Compressed Gas Cylinder Storage and Handling." *Cornell University Office of Environmental Health and Safety*, Nov. 7, 2005, http://weill.cornell.edu/ehs/static_local/pdfs/Compressed_Gases.pdf (Accessed Aug. 20, 2012).

"Confederate Steam Gun." *Mythbusters* TV science series, Episode 93, Discovery Channel, Dec. 5, 2007, via *Mythbusters Results*, www.mythbustersresults.com/episode93 (accessed Nov. 11, 2011).

Cox, Vicki. "Having a Blast." *American Profile*, Oct. 29, 2006.

Craig, Dudley P. *Steam Power and Internal Combustion Engines.* New York: McGraw-Hill, Inc., 1931.

"Crash at Crush" *Waco Heritage & History*, Vol. 8, No. 3, Fall 1977, quoted in "Train Crash at Crush." http://buckcreek.tripod.com/traincrash.html (accessed Oct. 10, 2006).

Croman, John. "State Fines Regions in Worker Death." *KARE NBC-TV 11* (Minneapolis-St. Paul), May 2, 2005, www.kare11.com/news/news_article.aspx?storyid=93343 (accessed Nov. 16, 2011).

Crow, Kirsten. "Magic Bullets." *Laredo Morning Times*, Jan. 20, 2008.

Cummins, Lyle. *Internal Fire: The History of the Internal Combustion Engine.* Wilsonville, Ore.: Carnot Press, 1976.

Dell & Schaefer Law Firm, "Miami Woman Hurt by Pressure Cooker Dies, Reported by Miami Accident Lawyer," news release, June 7, 2011, http://southflinjury.com?p=1120 (accessed Aug. 6, 2012).

Dickinson, Charles S. "Winans Steam Gun." *Harper's Weekly*, May 25, 1861.

Diener, Terry. "Disaster, 114 Years Ago This Month." *The Daily Item* online newspaper, Oct. 18, 2010.

Dingus, Anne. "The New London School Explosion." *Texas Monthly*, Mar. 2001, www.newlondonschool.org/Articles13.htm (accessed Jan. 16, 2012).

"Earthquake Machine." *Mythbusters* TV science series, Episode 60, Discovery Channel, Aug, 30, 2006, via *Mythbusters Results*, www.mythbustersresults.com/episode60 (accessed Aug. 15, 2012).

"East Ohio Gas Company Explosion." *Ohio History Central*, July 1, 2005, www.ohio historycentral.org/ (accessed Mar. 19, 2009).

"East Ohio Gas Co. Explosion and Fire." *The Encyclopedia of Cleveland History*, Case Western Reserve University, Mar. 27, 1998, http://ech.cwru.edu/ech-cgi/article.pl?id=EOGCEAF (accessed Aug. 20, 2012).

Echevveria, Emiliano and Walter Rice, PhD. "When Stationary Steam Engines Powered Cable." *The Cable Car Home Page*, Jan. 1, 2005, www.cable-car-guy.com/html/ccsfwhensteam.html (accessed June 17, 2009).

Edwards, Owen. "Inventive Abe." *Smithsonian Magazine*, Oct. 2006, www.smithsonianmag .com/history-archaeology/object-oct06.html (accessed May 25, 2011).

Emerson, Jason. "A Man of Considerable Mechanical Genius: Abraham Lincoln Was the Only U.S. President to Hold a Patent." *Invention & Technology*, Winter 2009: 10–13.

Emigh, Margie L. "Nine Boiler Accidents That Changed the Way We Live." *National Board BULLETIN*, Summer 2003: 20–25.

Enkoji, M. S. "Man Pushes Paintball Safety Warnings After Wife's Death." *Sacramento Bee*, Oct. 8, 2007, www.scrippsnews.com/node/27409 (accessed Dec. 8, 2010).

"The Exploding Beer Keg of Death." *CNHI News Service*, Mar. 10, 2007.

"Exploding Keg Kills Host," Associated Press. *New York Times*, August 4, 1981. *www.nytimes.com/1981/08/04/us/exploding-keg-kills-host.html (accessed Sept. 4, 2008).*

"Explosion Kills 1 at Connecticut Party." Associated Press. FoxNews.com, October 23, 2006. www .foxnews.com/wires/2006Oct23/0,4670,BRFPartyExplosion,00.html (accessed Aug. 26, 2008).

Fein, A., A. Leff, and P.C. Hopewell. "Pathophysiology and Management of the Complications Resulting from Fire and the Inhaled Products of Combustion." *Critical Care Medicine* 8 (Feb. 1980): 94–98. National Institutes of Health PubMed, www.ncbi.nlm.nih.gov/pubmed/7353393 (accessed Aug. 18, 2012).

Fisher, Karen. "Steam Engine, Alexandria, 100 CE." Smith College Museum of Ancient Inventions, www.smith.edu/hsc/museum/ancient_inventions/steamengine2.html (accessed Aug. 12, 2008).

Fitzgerald, H.J. "World's First Steam-Driven Airplane," *Popular Science Monthly*, Vol. 123, No. 1, July 1933: 9–11, 92. Via *Modern Mechanix* blog, Nov. 3, 2001, http://blog.modernmechanix.com/worlds-first-steam-driven-airplane/ (accessed Aug. 21, 2012).

Flam, Lisa. "With Arm Stuck in Boiler, Man Contemplated Suicide." *AOL News*, June 22, 2010.

Foster, Peter. "'Exploding' Watermelons Hit Farmers in China." *Telegraph* (UK), May 17, 2011, www.telegraph.co.uk/news/newstopics/howaboutthat/8518547/Exploding-watermelons-hit-farmers-in-China.html (accessed Oct. 3, 2011).

Gaylord, Tom. "What About CO_2?" 2003, *PyramidAir.com*, www.pyramydair.com/site/articles/co2 (accessed Sept. 3, 2008).

"GenDisasters Ship Disasters." *GenDisasters.com*, www3.gendisasters.com/taxonomy_menu/3/70 (accessed June 18, 2009).

"Ghost Hunt of the Kimo Theater, Albuquerque, N.M." *Southwest Ghost Hunters Association*, www.sgha.net/nm/albq/kimo.html (accessed June 11, 2009).

Gray, Charlotte. *Reluctant Genius: The Passionate Life and Inventive Mind of Alexander Graham Bell*. New York: Arcade Publishing, 2006.

Gregory, Andrew. "Scalding Horror as Coffee Machine Explodes in Sainsbury's Supermarket Cafe." *Daily Mirror* (UK) Online, Sept. 15, 2010.

Grigsby, John. "Gas-Storage Precautions Followed 1944 Explosion." *Toledo Blade*, Dec. 2, 1984: B1.

Hamilton, Allen Lee. "Crash at Crush." *Handbook of Texas Online*, http://www.tshaonline.org/handbook/online/articles/llc01 (accessed Oct. 10, 2006).

The Handbook of Texas Online. Texas State Historical Association, www.tshaonline.org/handbook/online (accessed Oct. 10, 2006).

Harris, Mark. *Grave Matters: A Journey Through the Modern Funeral Industry to a Natural Way of Burial*. New York: Scribner, 2007. (Accessed via *Google Books*, Jan. 17, 2012).

Hartwell, R.M. *The Industrial Revolution and Economic Growth*. Methuen and Company, 1971.

Hathaway, Peter B, Eric J. Stern, Richard C. Harruff, and David M. Heimbach. "Steam Inhalation Causing Delayed Airway Occlusion." *American Journal of Roentgenology*, 1996: 166–322.

"The History of Human Cannonballs: A Brief Timeline," Ontario Science Centre.

"How Cable Cars Work." *San Francisco Cable Car Museum*, www.cablecarmuseum.org/mechanical.html (accessed Nov. 17, 2008).

Hulse, David K. *The Early Development of the Steam Engine*. Leamington, UK, TEE Publishing, Ltd., 1999.

"Human Cannonballs," *New Scientist Invention Blog*, May 15, 2006, www.newscientist.com/blog/invention/2006/05/human-cannonballs_15.html (accessed Mar. 11, 2011).

Hurley, David. "Roof Lifted off in 'Freak' Explosion." *Portsmouth* (UK) *News*, Mar. 19, 2010, www.portsmouth.co.uk/newshome/Roof-lifted-off—in.6167902.jp (accessed Oct. 15, 2010).

"Idaho Man Dies When Explosive Beer Keg Lets Go." Associated Press. *Seattle Times*, Jan. 31, 1992, http://community.seattletimes.nwsource.com/archive/?date=19920131&slug=1473198 (accessed Sept. 8, 2008).

Jeter, Stephanie. "Assembling Pieces of the Past." *Tyler* (Tex.) *Morning Telegraph*, March 18, 2007.

Jha, Alok. "Is Being Fired out of a Cannon Dangerous?" *The Guardian* (UK), Sept. 1, 2005, www.guardian.co.uk/science/2005/sep/01/thisweekssciencequestions (accessed Jan. 28, 2008).

"Keg Explodes, Killing Man at Birthday Party." Associated Press. *The Stars and Stripes*, Nov. 1, 1988: 7.

"Key Dates in Fire History." National Fire Protection Association, www.nfpa.org (accessed Sept. 30, 2009).

Knox, Robert. "The Untold Story of Boston's Great Molasses Flood." *Boston Globe*, Jan. 11, 2004, www.boston.com/ae/books/articles/2004/01/11/the_untold_story_of_bostons_great_molasses_flood/ (accessed Feb. 2, 2007).

Lahanas, Michael. "Heron of Alexandria." www.mlahanas.de/Greeks/HeronAlexandria.htm (accessed Oct. 28, 2011).

The Last Voyage, film, directed by Andrew L. Stone. (1960; Metro-Goldwyn-Mayer).

Lienhard, John H. "High-Pressure Steam Engines." Radio series transcript. University of Houston College of Engineering, www.uh.edu/engines/epi1572.htm (accessed Sept. 14, 2011).

"Lightning Kills Diver Off Florida Coast." *San Francisco Examiner*, July 23, 2007.

"LNG Explosion Levels One Square Mile (Cleveland 1944)." *ERM Risk and Safety Blog*, Mar. 29, 2010, www.ircrisk.com/blognet/?tag=/1944 (accessed Aug. 20, 2012).

"The London Beer Flood of 1814." *h2g2.com*, www.h2g2.com/approved_entry/A42129876 (accessed July 19, 2011).

"The London Beer Flood of 1814." *Historical and Regency Romance UK*, March 7, 2010, www.historicalromanceuk.blogspot.com/2010/03/london-beer-flood-of-1814.html (accessed Sept. 7, 2010).

Mackinnon, Francis. "Some Terrific Explosions." *The Harmsworth Monthly Pictorial Magazine*, Vol. 2, Feb. 1899: 541–544, via *Google Books* (accessed Aug. 21, 2012).

"Man Dies After Explosion at Party." Associated Press. *Hartford Courant*, Oct. 23, 2006.

"Man Turns to 'Human Balloon' After Falling on Air Hose," *FoxNews.com*, May 25, 2011, www.foxnews.com/health/2011/05/25/man-turns-human-balloon-falling-air-hose/ (accessed May 25, 2011).

Manning, Michael, Sarah Smith, and Eric Chernoff. "Basics of Bootleggin' and Moonshinin'." *North Carolina Moonshine*, Nov. 25, 1997. www.ibiblio.org/moonshine/make/basics.html (accessed Sept. 30, 2009).

Marks, Peter. "OUR TOWNS; Target Practice for Life as a Human Cannonball." *New York Times*, Aug. 31, 1993, http://www.nytimes.com/1993/08/31/nyregion/our-towns-target-practice-for-life-as-human-cannonball.html (accessed Sept. 30, 2009).

Mason, John. "The Molasses Disaster of January 15, 1919." *Yankee Magazine*, Jan. 1965: 52–53, 109–111, via Eric Postpischil's Molasses Disaster Pages, http://edp.org/molyank.htm (accessed Feb. 16, 2007).

May, Irvin M. Jr. "New London School Explosion." *Handbook of Texas Online*, www.tshaonline.org/handbook/online/articles/yqn01 (accessed Jan. 16, 2012).

May, Thomas. "Oxygen-Filled Balloons Blow Up Car." *MyFoxAL.com* (WBRC Fox-6 TV, Birmingham, Ala.), Mar. 23, 2011, www.myfoxal.com/Global/story.asp?S=14308522 (accessed Mar. 25, 2011).

McGeehan, Patrick. "See Ya! Demolishing Yankee Stadium, Carefully." *New York Times* City Room blog, Sept. 5, 2008, http://cityroom.blogs.nytimes.com/2008/09/05/see-ya-demolishing-yankee-stadium-carefully/ (accessed Sept. 14, 2011).

Megyesy, Eugene F. *Pressure Vessel Handbook, 14th ed*. Oklahoma City: PV Publishing, Inc., 2008.

Merrell, Don. "Don Merrell Poems and Limericks." Safetycal, Inc., www.safetycal.com/store/safety_products/don_merrell_poems (accessed April 4, 2011).

"Miami Woman Dies After Exploding Pressure Cooker Severs Leg." *Fox News Latino*, June 6, 2011, www.latino.foxnews.com/latino/news/2011/06/08/miami-woman-dies-after-exploding-pressure-cooker-severs-her-leg/ (accessed Nov. 11, 2011).

Michner, Joerg. "Crematorium to Help Heat Homes in Swedish Town." *The Telegraph* (UK), Dec. 20, 2008, www.telegraph.co.uk/news/worldnews/europe/sweden/3869095/Crematorium-to-help-heat-homes-in-Swedish-town.html (accessed Aug. 21, 2012).

Mikkelson, Barbara. "Solitaire." *Snopes.com*, Jan. 19, 2007, www.snopes.com/horrors/freakish/kogut.asp (accessed July 19, 2011).

Murray, Tom. "Hidden Hazards of Circus Thrillers." *Modern Mechanix Hobbies and Inventions*, July 1936: 50–51, 131, via Modern Mechanix blog, July 15, 2007, http://blog.modernmechanix.com/2007/07/15/hidden-hazards-of-circus-thrillers/.

"NASA Spinoffs: Fact or Myth?" *Invention & Technology*, Fall 2008, Vol. 23, Issue 3: 38–39.

National Academy of Sciences — National Research Council. "Human Acceleration Studies." Publication 913, 1961, www.dtic.mil/dtic/tr/fulltext/u2/266076.pdf (accessed Sept. 13, 2011).

National Aeronautics and Space Administration (NASA). "Exploring Aeronautics: Aircraft Types Educator Guide." 2003, http://quest.arc.nasa.gov/projects/aero/ExploringAero/PDF/Aircraft%20Types.pdf (accessed Sept. 13, 2011).

____. "Brief History of Rockets." Feb. 12, 2010, www.grc.nasa.gov/WWW/k-12/TRC/Rockets/history_of_rockets.html (accessed June 14, 2011).

New London Texas School Explosion, www.newlondonschool.org (accessed Nov. 7, 2011).

"New Milford, Conn.: Explosion at Party Kills Man." *Associated Press*, Oct. 23, 2006, via *New York Times*, http://www.nytimes.com/2006/10/23/nyregion/23mbrfs-004.html?_r=1 (accessed Oct. 23, 2006).

"No Criminal Charges to be Filed in Fatal Keg Blast." *Republican-American* (Waterbury, Conn.), Mar. 24, 2007. www.NewsLibrary.com/ (accessed Aug. 26, 2008).

"Nothing Says 'Hello' Like Disaster on a Postcard." *National Board BULLETIN*, Fall 2004: 10–17.

Novak, Matt. "The Year 2011, According to Thomas Edison in 1911." *io9.com*, Jan. 23, 2011, http://io9.com/5741015/the-year-according-to-thomas-edison-in-1911 (accessed Aug. 21, 2012).

O'Brien, Dave. "2 Dead in Suffield Blast." *Record-Courier* (Ravenna, Ohio), Mar. 3, 2011, www.recordpub.com/news/article/4991674 (accessed March 3, 2011).

"Oct. 20, 1944: Natural Gas Explosions Rock Cleveland." *History.com*, www.history.com/this-day-in-history/natural-gas-explosions-rock-cleveland (accessed Aug. 20, 2012).

"Ohio House Explodes, 2 People Unaccounted for." *Columbus Dispatch*, Mar. 2, 2011, www.dispatch.com/live/content/local_news/stories/2011/03/02/02-house-explodes.html (accessed Mar. 2, 2011).

Olesen, Alexa. "Fields of Watermelon Burst in China Farm Fiasco," *Associated Press*, May 17, 2011. http://cnsnews.com/news/article/fields-watermelon-burst-china-farm-fiasco (accessed Aug. 18, 2012).

O'Malley, Julia. "Body Discovered in Duct After School Staff Detects Odor." Anchorage Daily News, *Free Republic*, Sept. 6, 2007, www.freerepublic.com/focus/f-news/1892083/posts (accessed June 11, 2009).

Park, Edwards. "Without Warning, Molasses in January Surged over Boston." *Smithsonian*, Nov. 1983: 213–230, via *Eric Postpischil's Molasses Disaster Pages*, http://edp.org/molpark.htm (accessed Feb. 16, 2007).

Perry, R.H., D.W. Green and J.O. Maloney, "Chapter 6: Psychrometrics," in *Perry's Chemical Engineers' Handbook, 6th ed.* New York: McGraw Hill, 1984.

"Popcorn: Ingrained in America's Agricultural History." *National Agricultural Library Special Collections*, Feb. 15, 2002, www.nal.usda.gov/speccoll/images1/popcorn.html (accessed June 6, 2012).

Potter, Jerry O. *The Sultana Tragedy: America's Greatest Maritime Disaster.* Gretna, La.: Pelican Publishing Co., 1997.

Puleo, Stephen. Dark Tide: *The Great Boston Molasses Flood of 1919.* Boston: Beacon Press, 2003.

Pyatt, Jamie. "Coffee Bomb." *The Sun* (UK) *Online*, Sept. 15, 2010, www.thesun.co.uk/sol/homepage/news/3138214/Coffee-machine-blast-hurts-fifteen.html (accessed Sept. 20, 2010).

Radinsky, Mike. "Huge Civil War-Era Gun A Curious Piece of Elkridge History." Nov. 28, 2010, *Elkridge Patch* Web site, http://elkridge.patch.com/articles/huge-civil-war-era-gun-a-curious-piece-of-elkridge-history.

"Red Rag to a Bull." *Mythbusters* TV science series, Episode 85, Discovery Channel, Aug, 22, 2007, via *Mythbusters Results*, www.mythbustersresults.com/episode85 (accessed Sept. 30, 2009).

Reece, Kevin. "Kent Man Killed by Exploding Lava Lamp." *KOMO News* (Seattle TV-4), Nov. 29, 2004, www.komonews.com/news/archive/4139111.html (accessed Nov. 15, 2011).

"Regions Fined in Scalding Incident." Antiscald, Inc., http://antiscald.com/prevention/scaldnews/fined.php (accessed Nov. 11, 2011).

Rosen, William. *The Most Powerful Idea in the World.* New York: Random House, 2010.

Ryan, Donald E. Jr. "The Airship's Potential for Intertheater and Intratheater Airlift." Academic thesis, School of Advanced Airpower Studies, 1992, www.dtic.mil/dtic/tr/fulltext/u2/a425995.pdf (accessed Sept. 13, 2011).

Ryan, Terri Jo. "'Crash at Crush' A Calamity Beyond Hypesters *[sic]* Dreams." *Waco Tribune-Herald*, Sept. 15, 2006, via Waco History Project, www.wacohistoryproject.org/Moments/CrashCrush.htm (accessed Jan. 4, 2010).

"Sainsbury's Finds Coffee Machine Explosion Fault," *BBC News Online*, Sept. 15, 2010. http://www.bbc.co.uk/news/uk-england-hampshire-11320258 (accessed Sept. 20, 2010).

Sanders, Roy E. *Chemical Process Safety: Learning from Case Histories.* Burlington, Mass: Elsevier Butterworth-Heinemann, 2005.

Sandukas, Gregory P. "Gently Down the Stream: How Exploding Steamboat Boilers in the 19th Century Ignited Federal Public Welfare Regulation [Redacted Version]." Third-year paper, Harvard Law School, April 30, 2002, http://leda.law.harvard.edu/leda/data/530/Sandukas_redacted.pdf (accessed Aug. 22, 2012).

Santos, Melissa. "The Human Cannonball Only Makes Flight Look Easy." *The* (Tacoma-Seattle) *News Tribune*, Sept. 21, 2008, via *HeraldNet* (Everett, Wash.), http://heraldnet.com/article/20080921/NEWS03/709219887 (accessed Sept. 22, 2008).

Shatzman, Israel. *Illustrated Encyclopedia of the Classical World.* New York: Harper & Row, 1975: 234.

Smith, H. Oram. "The London, Texas, School Disaster." *National Fire Protection Association Quarterly* 30, No. 4 (April 1937): 299–311.

Spitzner, Kirk, ed. *Twaintimes – A Time Line of Events in the Life of Samuel Langhorne Clemens.* Nov. 16, 1999, www.twaintimes.net/page1.html (accessed June 8, 2009).

____. "Care and Feeding of a Steam Boiler." *Twaintimes – A Time Line of Events in the Life of Samuel Langhorne Clemens.* Nov. 16, 1999, www.twaintimes.net/boat/sbpage2.html (accessed June 8, 2009).

____. "A Short History of Steam Engines." *Twaintimes – A Time Line of Events in the Life of Samuel Langhorne Clemens.* Nov. 16, 1999, www.twaintimes.net/boat/sbpage4.html (accessed June 8, 2009).

"The Stanley Steam Engine." *StanleyMotorCarriage.com*, www.stanleymotorcarriage.com/SteamEngine/SteamEngineGeneral.htm (accessed Aug. 16, 2010).

"The Stanley Steamer, Why the Fascination?" *StanleyMotorCarriage.com*, www.stanleymotorcarriage.com/GeneralTechnical/GeneralInfo.htm (accessed Mar. 11, 2011).

"Steam." *Webster's Online Dictionary*, www.websters-online-dictionary.org/definition/steam (accessed May 28, 2008).

"Steamboats 1811–61." *Steamboat Times*, 2007, http://steamboattimes.com/steamboats_1811~61_p1.html (accessed Sept. 29, 2011).

"Steam Car Technology: A Labor of Love." *National Board BULLETIN*, Fall 2003: 8–18.

Stowers, Carlton. "Today A Generation Died." *Dallas Observer*, Feb. 21, 2002, via www.therainwatercollection.com/reference/1580.pdf (accessed Jan. 10, 2007).

"Strange-Looking Car Leads to Explosives Charges." ABC-7 TV News (Denver) via *TheDenverChannel.com*, www.thedenverchannel.com/news/6790966/detail.html (accessed Sept. 8, 2008).

"Study: Steam Hybrids Using Waste Heat Recovery Could Reduce Fuel Consumption up to 31.7%." *Green Car Congress*, April 28, 2008, www.greencarcongress.com/2008/04/study-steam-hyb.html (accessed Oct. 29, 2008).

Stultz, S.C, and J.B. Kitto. "Introduction to Steam," in *Steam: Its Generation and Use, 41st ed.* Charlotte, N.C.: Babcock & Wilcox, 2005.

"Suits Filed in Fatal Keg Blast." *Republican-American* (Waterbury, Conn.), Jan. 18, 2007. www.NewsLibrary.com/ (accessed Aug. 26, 2008).

"Sylvester Roper." *American Motorcycle Association Hall of Fame*. www.motorcyclemuseum.org/halloffame/detail.aspx?RacerID=264 (accessed Sept. 11, 2011).

Teixidor, H.S., G. Novick, E. Rubin. "Pulmonary Complications in Burn Patients." *Journal of the Canadian Association of Radiologists*, 1983: 264–270.

Thurston, Robert Henry. *A History of the Growth of the Steam Engine*. London: Keegan Paul and Trench, 1883: 21–22.

____. *Steam-Boiler Explosions, in Theory and in Practice*. New York: J. Wiley & Sons, 1894.

Toynbee, Arnold. *Lectures on the Industrial Revolution of the Eighteenth Century in England*. Paperback edition. Whitefish, Mont.: Kessinger Publishing, 2004. Originally published 1884.

Tucker, Farrell L. "The Great Locomotive Explosion: A Socio-Historical Examination of a Tragedy." University of Texas at San Antonio Website, http://colfa.utsa.edu/users/jreynolds/Tucker/exp1.html (accessed Aug. 16, 2010).

"Turbine." *Encyclopaedia Britannica Online*, www.britannica.com/EBchecked/topic/609552/turbine (accessed Aug. 15, 2008).

U.S. v. Gagnon, 951 F.2d 350 (6th Cir. 1991). Dec. 27, 1991, via *Fastcase.com* (accessed Aug. 26, 2008).

Vaeth, J. Gordon. *They Sailed the Skies: U.S. Navy Balloons and the Airship Program*. Annapolis, Md.: Naval Institute Press, 2005: 95–96.

Varrasi, John. "The True Harnessing of Steam." *Mechanical Engineering Magazine*, Jan. 2005, www.asme.org/kb/news---articles/articles/boilers/the-true-harnessing-of-steam (accessed Aug. 21, 2012).

Walker, Duncan. "Head for Heights Required." *BBC News*, Aug. 22, 2005.

Watts, Jonathan. "Exploding Watermelons Put Spotlight on Chinese Farming Practices." *The Guardian* (UK), May 17, 2011. www.guardian.co.uk/world/2011/may/17/exploding-watermelons-chinese-farming (accessed Aug. 18, 2012).

"What Is the Most Dangerous Part of Being a Human Cannonball?" *Physics Central*, www.physicscentral.com/explore/poster-cannonball.cfm (accessed Mar. 11, 2011).

Whipps, Heather. "How the Steam Engine Changed the World." *LiveScience.com*, www.livescience.com/2612-steam-engine-changed-world.html (accessed Oct. 2, 2009).

Wilkins, Alasdair. "The Greek Engineer Who Invented the Steam Engine 2,000 Years Ago." Jan. 25, 2011, *io9.com* science and technology information site, http://io9.com/5742457/the-ancient-greek-hero-who-invented-the-steam-engine-cybernetics-and-vending-machines (accessed Aug. 20, 2012).

Wilkins, Robert. "Remembering St. Patrick's Day 1909." *Westmount Examiner* (Montreal), March 19, 2009.

Williams, Christopher. "Darpa Plots Emergency Man-Cannon." *The Register* (UK) *Online*, May 16, 2006, www.theregister.co.uk/2006/05/16/man_slinger/print.html (accessed Aug. 11, 2012).

"Woman's Leg Severed by Flying Pressure Cooker." *Local 10-TV* (Miami), May 19, 2011.

World Records Academy. "Smallest Steam Engine Made by Iqbal Ahmed Set *[sic]* World Record," news release, July 21, 2007, www.worldrecordacademy.com/smallest/smallest_steam_engine_made_by_Iqbal_Ahmed_set_world_record_70662.htm (accessed Sept. 14, 2011).

"World's Smallest Steam Engine Wieghs *[sic]* Less Than 2 Gms." *Hindustan Times* (New Delhi) July 21, 2007, www.hindustantimes.com/News-Feed/Science/World-s-smallest-steam-engine-wieghs-less-than-2-gms/article1-237948.aspx (accessed Sept. 14, 2011).

Yates, Donald. "The Vimo Ginger Beer Saga 1904 to 1090." *Bottles and Extras*, Winter 2004: 14–18.

Zibulewsky, Joseph, MD. "Defining Disaster: The Emergency Department Perspective." *Baylor University Proceedings, 2001*, 2001: 144–149. National Institutes of Health PubMed, www.pubmedcentral.nih.gov/articlerender.fcgi?artid=1291330 (accessed Mar 20, 2008).

INDEX